Urban Design Under Neoliberalism

This book discusses the status of urban design as a disciplinary field and as a practice under the current and pervasive neoliberal regime. The main argument is that urban design has been wholly reshaped by neoliberalism. In this transformation, it has become a discipline that has neglected its original ethos – designing good cities – aligning its theory and practice with the sole profit-oriented objectives typical of advanced capitalist societies. The book draws on Marxism-inspired scholars for a conceptual analysis of how neoliberalism influenced the emergence of urbanism and urban design. It looks specifically at how, in urbanism's everyday dimensions, it is possible to find examples of resistance and emancipation. Based on empirical evidence, archival resources, and immersion in the socio-spatial reality of Santiago de Chile, the book illustrates the way neoliberalism compromises urban designers' ethics and practices, and therefore how its theories become instrumental to the neoliberal transformation of urban society represented in contemporary urbanisms.

It will be a valuable resource for academics and students in the fields of architecture, urban studies, sociology, and geography.

Francisco Vergara Perucich is an urbanist at Universidad Central de Chile and Doctor in Development and Planning at University College London. Currently, he is Associate Professor at the Universidad de Las Américas, Chile, where he conducts research on neoliberal urbanisms and informal settlements in Latin America.

Routledge Focus on Urban Studies

Routledge Focus on Urban Studies offers a forum for cutting-edge research on a wide range of topics and issues within urban studies. The series aims to showcase the latest thinking on cities, from policy, governance, development, cultural and social topics, planning and design, as well as the latest analytical or theoretical innovations. The series also provides a forum for short topics aimed at specialized audiences and in-depth case studies that draw on a particular geographic locale. The format for the series is distinctive: each Focus is longer than a journal article and shorter than a traditional monograph.

Creating Modern Athens
A Capital between East and West
Denis Roubien

Urban Design Under Neoliberalism
Theorising from Santiago, Chile
Francisco Vergara Perucich

Urban Design Under Neoliberalism

Theorising from Santiago, Chile

Francisco Vergara Perucich

Routledge
Taylor & Francis Group

LONDON AND NEW YORK

First published 2019
by Routledge
2 Park Square, Milton Park, Abingdon, Oxon OX14 4RN

and by Routledge
605 Third Avenue, New York, NY 10017

First issued in paperback 2020

Routledge is an imprint of the Taylor & Francis Group, an informa business

British Library Cataloguing-in-Publication Data
A catalogue record for this book is available from the British Library

Library of Congress Cataloging-in-Publication Data
A catalog record for this book has been requested

ISBN 13: 978-0-367-72972-1 (pbk)
ISBN 13: 978-0-367-19574-8 (hbk)

Typeset in Times New Roman
by Apex CoVantage, LLC

We must resist becoming docile instruments of neoliberalism. We must struggle for change. There will be a moment in which this neoliberal trend will expire because it is implausible to sustain such adverse model against majorities and we must be prepared for impulsing that change. It is the only way to find professional satisfaction disregarding money.

—Miguel Lawner,
Chilean Architecture Awarded
Santiago, 8th April 2019.

Contents

Figures

Tables

About the author

Francisco Vergara Perucich is an architect and urbanist at Universidad Central de Chile, with an MA in Architecture from Pontificia Universidad Católica de Chile, an MSc in Building and Urban Design in Development, and a PhD from The Bartlett Development Planning Unit. His career has combined architectural practice and urban design projects in the Metropolitan Region of Santiago and academic work directing theoretical and practical modules at Universidad Central de Chile. In 2012, he moved to London and studied at The Bartlett Development Planning Unit. Over five years, he specialised in critical urban theory based on the Marxist approach developed by Henri Lefebvre. He gave to his career a particular emphasis in relation to the right to the city and the critique of the disciplinary field of urbanism as a dependent discipline from economics and policits, rejecting its original ethos: developing good cities for all. After finishing his PhD, he moved to Antofagasta, Chile, where he developed research on spatial inequality and also engaged with dispossessed local communities that are struggling for their right to the city, with special emphasis on Latin American migrants living in slums in the middle of one of the driest deserts in the planet. He contributes as a visiting lecturer at the Economics Department of Universidad Católica del Norte and is a full-time researcher at Universidad de Las Américas.

Acknowledgements

I wish to thank to my family for their love and patience throughout the process of working on this book and on the PhD that sustained most of the information here presented. Furthermore, I wish to thank Nadja, Julian, León, and Violeta, my personal pack, who encourage me to try to make better cities for everyone by discussing its current contradictions. Also, I would like to thank several colleagues who have supported me throughout my work on this book, not necessarily in any specific order – Julio Dávila, Catalina Ortiz, Cristian Olmos, Cristian Silva, Martin Arias, Miguel Atienza, Luis Miguel Rodrigo, Lisandro Roco, Caroline Newton, Julia Wesely, Anna Koledova, Andrea Cubides, Julia Hanse, Signe Lindberg, Nathan Mahaffey, Raffef Abdelrazek, Graham Perring, Ernesto Lopez Morales, Viviana Fernandez, Francisco Diaz, José Abasolo, Felix Reigada, Carlos Aguirre, Nicolas Verdejo, Sebastian Gray, Marcelo Reyes, Simon Zelestis, Miguel Lawner, Walter Imilan, Michael Lucas, Colin Marx, Harry Smith, Kisnaphol Wattanwanyoo, Ana Maria Huaita, Rodrigo Caimanque, and Karinna Fernández, among several other colleagues – for their fruitful conversations and discussions at different stages of the work on this book. I also want to extend these acknowledgements to thank all colleagues in the Architecture, Design, and Construction Faculty at Universidad de las Américas; my good colleagues and friends from Department of Economics, Universidad Católica del Norte; and my teachers and partners at The Bartlett Development Planning Unit, UCL. Furthermore, most of this research wouldn't have been possible without the support of Becas Chile scholarship grant No. 1859/2013 and the FONDECYT grant number 11180569, which provided the funding. I want to thank to my editors at Routledge, Faye Leerink and Ruth Anderson, for all their support during the submission, editing, and production process. Finally, thanks to my friend and tutor during the elaboration of this book: Professor Camillo Boano. He is my mentor with reference to urban design under a neoliberal scheme, and I hope he will enjoy reading this book as much as I enjoyed writing it.

1 Introduction

I start this book by stating that urban design has been completely subsumed by neoliberalism. Its original ethos of designing good cities for living has been neglected. Historically, successful stories of urban design involve cohesive socio-cultural values that shaped cities and maintained an equilibrium based on human values (Golany, 1995). It has been stated that "the discipline through which social aspirations can be realized physically" (Canniffe, 2006a, p. 1). Under neoliberalism, the disciplinary field of urban design aligned its ethos with a set of modes of production that aim to maximise the efficiency of land markets and facilitate financial transactions of urban spaces.

This statement requires further discussion, and so throughout these pages I discuss my investigation of the phenomenon of urban design under neoliberalism, focussing on the ways in which urban designers participate in neoliberal contexts as instrumental actors by commodifying urbanisation processes and therefore essentially undermining the very possibility of developing good cities. As a preview of the conclusions of this research, I state that to contribute to developing better cities, urban design under neoliberalism must be rejected.

The idea of revolution as employed in this book has nothing to do with social violence. Rather, it involves a change in social relations through a deep transformation in the disciplinary field of urban design. For Henri Lefebvre, the concept of the revolution is used analytically to understand reality from a critical standpoint, not only to comprehend social phenomena but also to transform it through praxis (Lefebvre, 2003). In the context of this book, the term "revolution" implies an urban strategy represented by a set of theoretical reflections on urban design under neoliberalism. Through this process, the research presented throughout this work aims to illuminate the contradictions that occupy the practice, theory, and ethics of urban design under neoliberalism.

In summary, the term "revolution", in this research, refers to the possibility of replacing the current urban design theories and practices under neoliberalism with new methods of designing cities. In discussing such a revolution, I have constructed a virtual object which will be disputed throughout this research: urban design under neoliberalism. This virtual object is dynamic. That is, it changes throughout this book as findings and reflections are adopted into its theorisation. At one point in the book, urban design under neoliberalism is described as the deception of a supposed reality, while in another section it is defined as the speculative future anticipated as an object of analysis.

A virtual object is an analytical and methodological instrument, elaborated by Henri Lefebvre in *The Urban Revolution* (Lefebvre, 2003), which takes preliminary information from reality and uses that information to mentally construct a possible future. This future has no clear positive or negative connotations. Instead, it is a provocative reflection of what reality may be in the future based on the development of certain social relations. In a way, a virtual object is a speculative assumption. Therefore, the actual existence of urban design under neoliberalism is real but cannot be tested unless its true nature is unveiled and a future is examined in which the totality of this virtual object has become a reality.

Urban design under neoliberalism is a virtual object because it contains the possibility of a future based on preliminary reflections that evolve throughout the text as new historical, practical, and ethical reflections that are incorporated into its theorisation. Specifically, the concept of the virtual object is useful for setting up a series of characteristics of urban design practice which the neoliberal project has transformed and which progresses as the research advances. Thus, the virtual object is used to trace the potential trajectory of observed phenomena because it chronologically organises the book, moving from the general (the overall context of Chilean urban design) to the specific (urban design practitioners' everyday practices). In other words, this book starts with initial assumptions about this practice under a neoliberal regime and moves to a series of contextualised theoretical reflections for a specific case (i.e., Santiago's urban design decision environment). Through understanding the complexities of urban design under neoliberalism, its subversion becomes a way to advance building strategies. In summary, urban design under neoliberalism extravagantly favours profit-oriented methods for shaping the city for the benefit of a select group of individuals while completely disregarding the possibility of creating better spaces for the majority.

The work of urban designers is based on their ability to imagine different futures, and this work is constrained by the neoliberal framework. The profit-oriented scheme that prevails in the neoliberal context fosters a homogenised vision of cities and their aesthetics. This is because the use of previously tested models of urban products ensures their economic success. Any attempt to define spaces that goes beyond profit margins is discarded. The efficiency of and revenue generated by investments in urban areas are the main criteria for creating innovative solutions to problems in public spaces. Hence, under neoliberalism, urban design must consider not only the feasibility of certain projects but their profitability as well. This research tests these issues by enquiring into the disciplinary field of urban design in a neoliberal context.

The virtual object of urban design under neoliberalism discussed presently emerged from a review of the extensive theorisation of several authors on the relationship between neoliberalism and the disciplinary field of urban design (Araabi, 2017; Langhorst, 2015, Sorkin, 2013; Cuthbert, 2011; Goonewardena, 2011; Gunder, 2011; Short, 2006; Carmona, 2001). Although multiple definitions of this relationship have been offered, it can be asserted that they vary depending on context and the idea that neoliberal urban design may be used as a category of analysis. As a descriptive category, urban design under neoliberalism depicts spatial, temporal, material, and discursive urban design practices as means of developing urban spaces for the purpose of raising capital (Boano & Vergara-Perucich, 2017). To spur the advancement of urban design towards the creation of good cities, this book provides a theoretical reflection on the intellectual framework of urban design which has been widely criticised for being undetermined, soft, broad, and irrelevant (Boano, 2017; Carmona, 2014a; Cuthbert, 2006a). These critiques allude to the importance of theorising about urban design given the relevance of the public space in the formation of a more collective society. The under-theorised status of urban design facilitates the reproduction of neoliberalism, the consequences of which vary according to the city's context.

I employ a case as a practice-oriented discipline to better illustrate the conflicts inherent in urban design under neoliberalism in the case of Santiago de Chile. This city's context provides empirical evidence that depicts the effects of neoliberalism in urban design. By examining this specific case, the findings of this book contribute to the literature by theorising about a phenomenon that is prevalent in a metropolis of the global south – an area where neoliberalism has reshaped cities for the sake of increasing capital, thereby fostering a sense that the development of a city is a financial investment rather than a social project.

Throughout this book, I have employed the methodological insights of Henri Lefebvre, a French philosopher and sociologist whose career contributed to the development of a holistic Marxist understanding of society which was developed in more than 60 books and an impressive number of papers. Along with his vast work, Lefebvre developed diverse approaches to solving urban problems. Although he primarily developed theories, he constantly proposed methodologies for critically diagnosing urban realities. Among his most spatially oriented methodological contributions are *The Right to the City* (1968), a political approach to urban planning; *The Urban Revolution* (2003 [1970]), an ethical questioning of urbanisms; *The Production of Space* (1974), a sociological interpretation of the role of space in society; and *Rhythmanalysis* (2004), an empirical examination of everyday practices in space.

The Urban Revolution offers a methodology for urban studies that emerged from a spatial interpretation of Marx's and Lefebvre's understanding of the complexities that urban society would face if urbanisation continued to be dominated by capitalism. The urban society is the successor of the industrial society; thus, according to Henri Lefebvre, we all live in the era of the urban society. For Lefebvre, urban design practice – like other urbanisms – is a blind field that has been subjugated to satisfy the objectives of capitalist development, which define economic growth as the main goal of human activities.

In reflecting on the work of Lefebvre, it is possible to visualise the processes of capital accumulation in the production of urban spaces (Harvey, 2012). In urban society, industrialisation left its essence as a capitalist engine, and consequently capital needed new ways of reproducing itself. Urban spaces are immobile and cannot be destroyed easily, making them excellent long-term fixed-income assets. Therefore, it is possible to channel large amounts of capital investments into spatial production. Space efficiently produces surplus value through its construction and through its profit when it is sold. Moreover, the processes of spatial production allow for the capitalist exploitation of labourers, first as a workforce in the construction of buildings and second as mortgagors when purchasing the space (in the case of housing) or using it (in the case of public facilities) through taxes. The cycles of urban production are stable, efficient, and apparently endless, given the constantly increasing number of people living in cities (Aravena, 2016). However, Lefebvre foretells a moment of crisis, which will occur when urbanisation reaches its limit, when the entire population moves to cities and the demand for space reaches its peak. Lefebvre labelled this stage in history "the critical zone" (2003).

Chile represents a pertinent subject of enquiry given that 90% of Chileans now live in cities (World Bank, 2016). Lefebvre states that the practices of organising and designing urban spaces in the critical zone require a transformation from a capitalist to a humanistic approach. "Theoretical knowledge can and must reveal the terrain, the foundation on which it resides: an ongoing social practice, an urban practice in the process of formation" (Lefebvre, 2003, p. 17). Although most people in Chile live in cities, only a few people decide how to conduct the processes of urban development. Therefore, most people have no influence on the form of the built environments in which they live.

Urbanisation processes are implemented to promote capitalism, which is causing many problems for most of urban society, including fragmentation, alienation, segregation, and an everyday life oriented towards producing wealth for just a few people. City inhabitants have become a profit-oriented group of individuals struggling for their everydayness. Hence, social relations have undergone a deep – and perhaps irreversible – transformation. A dominant class controls society through markets and urbanisation. Although the scenario seems apocalyptical, Lefebvre was optimistic about the potential of urban design. From his perspective, in planetary urbanisation, several cracks in capitalism (crises) would open up possibilities for igniting a revolution that would free urban practices from capitalism.

For this reason, I have assumed the challenge of unpacking contemporary urban design to reveal the importance of urban practices in the overthrow of capitalism. Therefore, in this book, I use Lefebvre's *The Urban Revolution* as a tool to shape the structure and methods of the entire research process. Three aspects explain the contribution of Lefebvre's book as a methodological piece of work. First, it is an object "designed" by Henri Lefebvre that employs a dialectical logic for each topic that is analysed, thus providing a different interpretation of Marx's work. By contrasting theory with practice and ideology with knowledge, Lefebvre untangles the relations between capitalism and urban development. Second, it is a historical work. *The Urban Revolution* refers to the period after the industrial revolution. During this period, society advanced towards total urbanisation led by the objectives of a capitalist hegemonic class. Hence, the role of disciplines related to spatial production is crucial, whether for facilitating the goals of capitalism or for providing alternatives. In this case, neoliberalism represents a deepening of the capitalist ideology criticised by Lefebvre. Accordingly, the present work enquires into the neoliberal era as a period in which political, economic, and social transformations changed the way urban design is practiced in Santiago. Third, *The Urban Revolution* is a manifesto for the possibility of revolutionising urban development practices so that they can resist and antagonise capitalism by changing the prevailing modes of spatial production. It is also a manifesto for producing a second urban revolution which is stripped of capitalistic ideas and embedded into social processes to create a democratic society. For this reason, the term "revolution", for Lefebvre, pertains to both a historical moment of change and a complete transformation of a practice.

In *The Urban Revolution*, Lefebvre refers to several examples of urban phenomena in Europe, especially in France. He implies that the study of these cases provides a corpus of social relations to review. Just as Marx focussed his analysis on the most advanced industrial society of his time (England), this book examines one of the most radical expressions of neoliberalism in the current world by observing the case of Santiago de Chile, a city in which the rule of free-market economics has mutated the overall structure of society. Santiago has an extensive set of neoliberal urban policies and practices whose spatial representations and historical developments contribute to our understanding of the origin of and logic behind urban design under neoliberalism.

For Henri Lefebvre, capitalism and urban practices are theoretically and empirically articulated in a co-dependent relationship by reproducing the hegemonic dominance of a social class. As will be explained in this book, an oligarchy controls society in Santiago in ways by which urban spaces are organised to maximise profit and defend the owners of private property.

As previously mentioned, in order to illustrate how urban design has been subsumed by neoliberalism, this book employs *The Urban Revolution* as a methodological resource, taking three specific research strategies from it:

- Transductive reasoning: an approach to research that combines inductive and deductive methods to contest a theoretical object in the form of a hypothesis about a future reality – a utopia (or dystopia) of a possible society. The theoretical object of interest in the present research is "urban design under neoliberalism".
- Levels and dimensions of analysis: intended to dissect urban design under neoliberalism defined as (i) a global level of analysis of the historical dimension of a theoretical object (in this research, specifically the ethics and practices of urban design in Santiago); (ii) a mixed level that unravels the implementation of urban design under neoliberalism and its effects on the city; and (iii) a private level focussed on the ethical dimensions of urban designers under neoliberalism.
- Spatial dialectics: a method that situates urban spaces at the centre of a dispute between different actors. Whoever controls a space also defines how the economics and politics within that space are developed. Therefore, spatial dialectics enquire into the relationships among societal groups when defining the urban form. Furthermore, in studying this relationship, spatial dialectics illuminate the contradictions and cracks emerging from contestations against the processes of urban production.

By using these methodological strategies, the present research values the contribution of *The Urban Revolution* to researching urban design. Thus, this book discusses a controversial debate on the relationship between spatial practices (urban design) and the politico-ideological projects of neoliberalism. If urban spaces are used by neoliberalists for their own reproduction, cities are the perfect battlefield for critically understanding neoliberalists' operations and its wrongdoings.

Santiago offers diverse examples of how neoliberalism is contested by spatial practices. Indeed, Santiago is the first city in the world in which neoliberalism was implemented; it was put in place in 1975 (Harvey, 2005; Klein, 2011). Therefore, Santiago constitutes an archaeological-like example for studying neoliberal urban practices. An active group of scholars has criticised the spatial consequences of neoliberalism in this city. Taking a more general approach, neoliberal urbanism has been studied as a set of methods and practices that characterise the spatial development of Santiago (Aguilar, Oliva, & Laclabere, 2016; Janoschka & Hidalgo, 2014; López & Meza, 2014; Rodriguez & Rodriguez, 2009). There is an abundance of research which explains how neoliberal urban policies have been used to perpetuate segregation and fragmentation in the city (Cociña, 2016; Daher, 1991; Gurovich, 2000; Hidalgo-Dattwyler, Alvarado-Peterson & Santana-Rivas, 2017; Vicuña, 2013). In the decision-making process for the design and formation of cities, significant advances have emerged from the research of Jorge Inzulza-Contardo (2011). As urban design is comprehended as an activity that organises cities for social activities, an analysis of the methods of urban design under neoliberalism is central to contesting neoliberalism. According to this logic, I explore how space in Santiago has been converted into a financial asset, an elemental resource for capital accumulation.

As you read this book, you will see how urban design in Santiago has emerged as a discipline controlled by a ruling class in order to promote spatialised social control by which urban development is manipulated for the sake of capital accumulation and reproduction. My aims are to elaborate and explore the complicity of urban design in Santiago in relation to the objectives of neoliberalism, to examine the profit generated and social control imposed by the ruling class, and to identify the faults of urban design under neoliberalism. Ultimately, I hope to help liberate urban design from neoliberalism.

Research design

Urban design under neoliberalism is a complex subject because it is constructed from political, economic, and spatial perspectives. It also creates multiple effects and may be observed from outside the built environment as well as within it with regard to the design of institutions, regulatory frameworks, the functioning of markets, and social relations. This research endeavours to study urban design under neoliberalism, specifically:

- It follows the methodological framework of Henri Lefebvre, research in the disciplinary field of urban design, and the generation of a heterodox Marxist approach for analysing the case of Santiago de Chile.
- It aims to produce a methodological reflection of urban design under a neoliberalist society (i.e., Santiago), focussing on its ethics, theories, and practices.

Consequently, in searching for the relationship between neoliberalism (as a complex ideological project and a set of political, social, and economic agendas) and urban design practices, the main research question is: What is urban design under neoliberalism?

This question demands a deep reflection on how urban designers work and the ethos urban design under neoliberalism. Urban development is vital for reproducing cycles of capital accumulation in space, which also spatialises social classes. Starting in 1961, Henri Lefebvre studied this condition and argued for changes among urban practitioners, demanding that spatial specialists play a more active role in changing capitalist urban development into a social production of space embedded in the interest of the majority and unfettered from the leashes of capitalist goals to capture surplus value and reproduce cycles of capital accumulation.

In *The Urban Revolution*, Lefebvre exposed the potential of urban design practitioners to catalyse social transformations (Lefebvre, 2003), and yet their role remains open to critique. Through my enquiry, I attempt to promote an understanding of the position of urban designers in the neoliberal ruling regime, how this ideology affects their practice, and, ultimately, the relationship between neoliberalism and the form of a city. Analysing the relationship between neoliberalism and urban design using transductive reasoning may illuminate its pitfalls and unveil its contradictions, thus providing insightful ideas about how to subvert neoliberalism from the practice of city making. In order to deepen this analysis, I analyse the discipline of urban design beyond its products, reaching the realms of the decision-making processes, its historical development, and the ethos of the practitioners who shape and make the city to reveal its complex production.

The very nature of urban design produces a series of rules and outcomes based on projective, imaginative, future-oriented, realistic, and complex understandings of how urban life can be improved. However, since neoliberalism dominates the political, economic, and social agendas, urban design has adopted the logic of a chrematistic approach to disciplines, pursuing profit and maximising investment, thus abandoning its ethos of designing a good city (Amin, 2002). Presently, a disciplinary field recognised as the spatial political economy has incorporated urban design (Cuthbert, 2006a), filling its theoretical gaps with appropriate knowledge for organising its ethos. The mongrel condition of urban design (Carmona, 2014a) has created the potential for total neoliberalisation. Thus, separating neoliberalism from urban design requires very deep theoretical reflections. In recording these reflections, I present references to a theoretical register throughout the book.

The virtual object as a hypothesis

A virtual object results from a deep reflection of the possibilities of the transformation of a current urban society in an undetermined future. In this case, my reflection is focussed on how cities are developed under a neoliberal regime and what would happen if total neoliberalism were to prevail over society as a whole. Based on transductive reasoning, the hypothesis of this research is that urban design has been neoliberalised; in this thesis, this state of affairs is referred to as "urban design under neoliberalism". This means that an economic theory and an ideological project have penetrated all structures of urban life, making profit and money the primary objectives of urban designers. This is revealed when applying a spatial dialectical analysis to the process of the production of urban space. The theory of neoliberalism emerged from monetarism, an economic theory developed by Milton Friedman in the 1950s which proposed that economies are more efficient when they have an increasing supply of money in the long term, as this condition ensures the circulation of money in markets and the broadening of people's purchasing capacity (Brunner & Meltzer, 1972). This political economy became a political project whose influence on diverse governments around the world was fuelled by the governments of the United States and the United Kingdom in the 1970s and 1980s (Harvey, 2005). "Urban neoliberalism refers to the interaction of processes of neoliberalisation and urbanisation and how such ideology is shaping and producing the form, the image and the life in the cities" (Boano & Vergara-Perucich, 2017, p. 10).

Neoliberalism was implemented in Chile by Augusto Pinochet's dictatorship in 1975. Since then, the Chilean state has become neoliberal, and urbanisation processes have suffered from the dogmatism of neoliberal implementation. Consequently, urban design practitioners altered their theories, ethics, and practice in order to survive neoliberalism. Given that the main goal of neoliberalism is to build a profit-oriented society, urban design (just like other creative disciplines) had to renounce imagination and utopian thinking in order to be profitable. Urban design became more concerned with developing economic solutions than with creating better urban spaces, thus creating a dialectical contradiction within the discipline: the good city and the profitable city are often not the same in practice. Therefore, the preference for a profitable city has subjugated the possibility of developing a good city. This change represents the neoliberalisation of urban design by adopting monetarism as its own epistemology.

My proposal for this virtual object assumes that the main ethical and practical contradiction of neoliberal urban design in Santiago is that urban designers are aware of the neoliberalisation of their discipline and its effects on society. I assume that urban designers know that their practice has been distorted to be profit-oriented; still, they continue using the same methodological apparatuses to shape city spaces and urban life because the decision-making power remains in the hands of an oligarchy. Hence, the theory behind this behaviour needs to be constructed according to the practice and mindset of urban design practitioners while observing the spatial outcomes of urban design under neoliberalism.

One of the outcomes of the neoliberal project has been the alienation of people; hence, urban designers are not yet capable of fighting to emancipate their practice from monetarist epistemologies. Instead, they deal with the urgencies of urban life (e.g., poverty, transportation, agglomeration), and they have no time for theorising their subjugated condition or for imagining a different future. In practice, urban designers struggle alone from a lack of organisation. In the meantime, they need to keep their jobs to survive. Fear, individualism, and the supremacy of profit-oriented practices are obstacles between urban designers' self-criticisms and the overthrow of urban design under neoliberalism. Therefore, the main strategy for ending the neoliberal regime must start with an organised collective discussion on building a broad social agreement (contract) capable of stopping profit-oriented urbanism and advancing towards people-oriented city designs.

Outline of the book

This book is organised using transductive reasoning, and therefore I have created a virtual object (urban design under neoliberalism) to encapsulate how neoliberalism influences the disciplinary practice of urban design. The construction of this virtual object ignites the research process because it raises questions regarding the design of a methodological strategy that reveals its contradictions and faults. From there, the book uses the case of Santiago to explore the materialisation of this virtual object advances in a real-world case.

As part of this thesis's rationale, the first two chapters illustrate how urban design under neoliberalism is represented in Santiago. Then, Chapters 3,

4, and 5 explain the relationship between urban design and neoliberalism based on the information gathered from fieldwork in Santiago. Finally, Chapter 6 reflects on the findings and outlines how urban design under neoliberalism might be overcome. A more detailed description of the structure of this book's remaining chapters follows.

Chapter 2 presents a literature review to problematise the relationship between the disciplinary field of urban design and neoliberalism and how this relationship can be studied in the case of Santiago de Chile. In this chapter, I present definitions and perspectives on urban design as a mongrel discipline whose aim is to produce good cities. Also, this chapter describes the political and ideological dimensions of neoliberalism with a focus on Chile and the relationship between these two dimensions regarding city-production processes. Furthermore, this chapter briefly introduces the main discussions presented in this book as to how neoliberalism has changed the city of Santiago.

Chapter 3 provides a global analysis that aims to present a baseline for contextualising and understanding the historical origins of Santiago's urban design discipline to more accurately perceive its neoliberalisation. This chapter explains when urban design first appeared as scientific urbanism and how this approach to this disciplinary field changed during the twentieth century until the irruption of neoliberalism, which occurred after the coup d'état led by Augusto Pinochet.

Chapter 4 illustrates some key practices that have emerged from the neoliberalisation of Santiago, focussing on the way these changes have influenced the disciplinary field of urban design. This chapter explores the spatial outcomes of the institutional transformations that generated the practice of urban design under neoliberalism. The chapter also provides a contextualised perspective of Santiago as a metropolis and an explanation of how it was transformed by neoliberal policies, such as the National Policy of Urban Development of 1979, the social housing programmes of the 1990s, and the strong development of public-private partnerships in urban development.

Chapter 5 is an exploration of the praxis and ethics of urban designers in Santiago who wish to unveil their own concerns, contradictions, and convictions about their disciplinary involvement under the neoliberal regime. Most of this chapter is based on interviews with urban design practitioners. This chapter reveals concrete revolutionary possibilities for transforming this disciplinary field, starting with a critical understanding of urban design in Santiago.

Chapter 6 develops a series of theoretical reflections towards building a theory of urban design under neoliberalism. This chapter serves as a conclusion and assessment of the findings. Furthermore, this chapter answers the research questions and provides a series of recommendations for formulating an agenda for further research on urban design in neoliberal contexts. In answering the research questions, this chapter introduces some key reflections that contribute to the creation of a theory that adequately explains urban design under neoliberalism and offers approaches for subverting it.

2 Discussing urban design and neoliberalism in Santiago

Introduction

This chapter presents the current discussions that serve to articulate neoliberalism with the disciplinary field of urban design and explains why Santiago is a suitable case for studying the relationship between these elements. By doing so, the chapter presents relevant discussions in literature about the way neoliberalism – as ideology and political project – has influenced the practice of urban design and briefly introduces how these features have been studied in Santiago. The analysis of the relationship between urban design as disciplinary field and neoliberalism is not particularly abundant in literature. However, the use of cities in study cases for explaining certain features of neoliberalism may be found in various publications and studies that serve for exploring the theoretical links between these two objects of study. So far, the case of Santiago and its neoliberal urban development has been mostly focussed on public policy and social consequences but not as much on how the space is designed under a neoliberal regime. This chapter presents the main discussions about urban design, neoliberalism, and introduces Santiago as a suitable case for this research.

Urban designer's ethos

The term "ethos" refers to the guiding beliefs that characterise the behaviour of a certain group of individuals. "Ethos" implies the observation of the conduct and habits of people seeking to accomplish certain goals. In this case, the urban designer's ethos refers to the set of beliefs and attitudes that inform the decisions of urban design practitioners as they shape the urban form.

The theoretical scope for outlining the ethos of urban design lacks clarity, which represents one of its values: being a mongrel disciplinary field that borrows methods and frameworks from other disciplines for studying and theorising the urban space, dealing with some of the most complex urban problems (Carmona, 2014b). Aseem Inam said that the mongrel nature of urban design creates a vague and ambiguous set of methods constructed by multiple disciplines (Inam, 2002). One of the challenges of urban design is how to advance from the mere prescription of solutions for cities to a more socially embedded production of ideas (Sorkin, 2013), which should reflect a theoretical production of its own practice (A. R. Cuthbert, 2011). Because they work in a materialistic discipline, urban designers make decisions that have lasting physical representations in space, which requires them to think about the discipline's theorisation in order to assume a more responsible way of doing cities. Theory may provide a set of principles to test and may improve through practicing city making. Specifically, the disciplinary field of urban design deals with the creation and shaping of public spaces and facilities and the organisation of private initiatives and should reflect social needs and define the way in which cities are developed.

Public spaces include streets, roads, and pavements; parks, plazas and squares; public and private housing; hospitals and educational facilities; transport and communication systems. All are part of the most elementary scope of urban design practice. For Jonathan Barnett, the role of urban design is to give a physical design to urban transformations, including landscape. In his definition, urban design deals with how the city looks in spaces between private properties, facilitating the creation of coherent spaces by providing quality spaces for urban life (Barnett, 1982). Mostly, urban design is practiced through "policies, programs and guidelines rather than by blueprints that specify shape and location in detail" (Kevin Lynch in George, 1997). From the early theorisations of the disciplinary field of urban design it has been framed as a practice that plays a relevant role by acting in the political and economic realm for defining norms and regulations that ensure the development of a good city.

These norms and regulations are embedded in an institutional framework (government, local authorities, companies, organised communities) that defines goals in relation to the urban space for social needs by creating a decision environment, as an invisible web (George, 1997) for guiding the process of shaping the city. Contemporary urban design is not necessarily directly involved in the design of urban objects, so in many cases it operates as a "second-order design" (George, 1997, p. 150), acting as ghost designers. Urban designers' expertise involves setting the criteria and principles for designing cities in order to enhance urban life, even if the private interests attempt to take advantage of urban processes. In my experience, urban design may contribute actively to resolving the conflicts of interest over the space between capital and communities and to finding better solution for both parties.

Urban designers also play a key role in real estate developments by defining the layouts of urban projects, analysing the regulations to find investment opportunities, and designing the public areas of private projects. Diverse architectural projects need urban designers as well to define public spaces, such as shopping malls, housing buildings, and suburbs. Hence, as counterpart to those urban designers working at public institutions, the private practitioner attempts to maximise the investments while at the same time creating good urban spaces. In public or private realms of the city, urban design's role is fundamental to the processes of the production of the commons, of the spaces where we meet as society and where the politics of encounters occurs. Therefore, the understanding of urban design complexities can serve as a spatial entry point to discuss how political-economic projects influence urban life.

Advancing in outlining the ethos of urban design, methodologies for studying it vary depending on the particular interpretations of the practitioner. For instance, Kevin Lynch explored the elements of space for categorising a city, Christopher Alexander studied the patterns that explain cities, and Bill Hillier looked at the fluxes that shape the space, but all three approaches are insightful and significant despite their differential methodologies. Urban design is a complex disciplinary field that cannot be subsumed under a totality that eliminates the richness of conflicts emerging when study the urban form (A. Cuthbert, 2007). Generalisations on urban design are discussable because methodologies are not fixed; each city is unique. A general theory of urban design explaining all forms of cities seems unreasonable. Theories of urban design do not seek a general theory of urban design; instead, they focus on its outcomes: wellbeing of inhabitants, fair distribution of resources in cities, good access to public transport, accessible leisure facilities, and affordable housing, to mention some examples of desirable urban design outcomes.

Rather than searching for a general theory of urban design, it is possible to explore the theoretical elaborations on these examples. These explorative approaches may lead towards the creation of an idealised urban environment in order to orient the decision-making processes for city making. This means that it could be more realistic to advance theories for explaining the effects of urban design practices rather than a theory on the spatial characteristics of spaces. A theory of urban design will be more universal when it is capable of setting the principles for understanding the decision making on urbanisation processes rather than explaining the outcomes of urban design practices.

Urban design is a disciplinary field whose impact on society is high because it determines the organisation of the city, so views on urban design should be weighed against the way people see their cities, their expectations, and their needs. Therefore, the aesthetics of spaces, the experimental practices, the economic effects of spatial transformations, the environmental value of projects, and the social preferences are more certain elements for judging the assertiveness of urban design approaches (Biddulph, 2012).

Urban design is an applied disciplinary field that confronts the use of social science methods with arts, discussing the imperious necessity of transforming research findings on urban form by either norms, regulations, designs or actions. The creative nature of this disciplinary field should not be excluded from the analysis of its ethos. Urban design as a mongrel disciplinary field synthesises its scientific-artistic feature under a set of theoretical reflections with empirical outcomes. Social science methods are employed for analysing reality and informing decisions and then for using political and artistic techniques in developing proposals.

In recent disciplinary categorisations, urban design has been situated as a disciplinary field that articulates different modes of producing and studying the built environment (geography, urban planning, urban studies, architecture) in order to situate public space at the centre of the discussions on how to improve urban life through spatial changes (Bentley, 1976; Carmona, 1998; Madanipour, 2006). If in the 1980s the idea of urban design was mostly associated with defining public spaces' form, today it also involves the political arena and research on how cities help to improve the life of people. Recently, urban design has become more holistic because public space has been conceptualised as a political arena for social struggle against the dominant neoliberal ideology (Brenner, Marcuse, & Mayer, 2012). Mike Biddulph (2012) contended that urban design professional practice encompasses the use of theoretical developments about the built environment at a range of different scales for designing forms that improve the character of the public realm, which makes of urban design a very political practice.

If we consider urban design as part of social theory, the study of processes or the production of spaces represents a social praxis driven by specific means and aiming to goals defined by a dominant class, which makes it as an appropriate discipline to be analysed under a Marxist framework (Cuthbert, 2007). Urban designers' tasks are embedded in the chain of production of spaces the urban society, extracting value from urban transformations and using the city as a resource for capital accumulation (Lefebvre, 1991b). Therefore, urban design may play a critical role in stripping urban processes from capitalism (Lefebvre, 2003). Urban designers may become agents capable of changing social relations working within the urban system (Biddulph, 2012, p. 3) regardless of the dominant ideology by infiltrating politics and igniting strategic transformations in society by reorganising cities. Matthew Carmona has developed an extensive work creating categories that illustrate the central conflicts that urban design faces in the contemporary development of cities. Throughout his vast literature on urban design, it is possible to summarise the way urban designers are responsible for specific spatial consequences that undermine the quality of cities, most of them related to an uncontrolled market force that ends by shaping spaces for the sake of capitalists' interests.

The importance of the market in shaping urban spaces has increased since the emergence of neoliberalism (Harvey, 2006). As part of the neoliberal transformations, state expenditures are re-directed from social goods – such as housing, education, healthcare, pensions – towards security measures controlled by market agents and towards the use of subsidies to the private sector to provide these services (Johnson, 2011). Neoliberalism weakens the public sector, which directly influences the importance of urban design in this ideological understanding of society. Under a neoliberal regime, urban design must help to attract capital investments or increase the economic efficiency of spatial investments, rather than producing reflective designs that actually represent people's needs and address urban problems (Langhorst, 2015). This condition is stressed by the globalised networks of cities in which diverse iconic works are constructed by the interests of the transnational capitalist class that employs the fame of architects (namely starchitects) to attract investments (Sklair, 2005, 2006). The economic development of cities in the globalised era requires significant investments in the urban space to ensure their growth (Gospodini, 2002). Urban design serves as a mean of economic development.

Urban design in neoliberal contexts is used for allocating investments to add value to urban areas. Good urban design raises capital and rent in the surroundings of interventions by more than 20%, thus increasing the possibilities of ensuring the selling of spaces, breaking into new markets for expanding businesses, and contributing to enhancing company image in communities (Carmona, 2001). Under neoliberalism, entire areas of the city are benefited by the production of high-quality urban developments, but less attractive spaces for capital are neglected. Despite being privately owned, most of the open spaces designed for catching value are publicly accessible spaces (Foroughmand Araabi, 2017). So, from the perspective of the provision of good spaces, one of the problems of urban design with neoliberal policies is its uneven distribution. In under-developed cities, the profit-oriented logic of urban design plus neoliberal policies undermine the possibility of allocating investments in deprived areas of the city where value can be affected (Harvey, 2012). Neoliberalism fosters the competition between products as a measure for optimising investments. Hence, the less competitive spaces receive less attention from the market, and their chances of receiving investments for improvements are reduced because it is riskier for capital. From the perspective of the urban designers' role in physically organising the city, this uneven logic of making areas of the city compete based on the profitability of development in each area hampers the possibility of developing just cities. It is the problem with supply and demand as the main decision-making criterion.

The neoliberal agenda has undercut the funding for urban design in the sense that the public sector could advance towards providing a good life through urban spaces (Short, 2006). Michael Gunder goes further and observes that the disciplinary field of urban design's lack of critical postures against the neoliberal agenda because contemporary urban design is, indeed, a creation of neoliberalism. Urban design adopted the economistic approach to society as a means to stress the role of built environment as a commodity (Gunder, 2011). This view echoes with interpretation of Henri Lefebvre, who stated that urban disciplines are masked instruments of capitalism (Lefebvre, 2003) that transform urbanisation into a mechanism for provoking cycles of capital accumulation through the production of space (Harvey, 2009; Lefebvre, 1991b). "Following on from this argument, when topics such as social justice, emancipatory design and gender are discussed under the titles of urban design, they cannot act otherwise than required by neoliberal forces" (Foroughmand Araabi, 2017, p. 4). The articulation between urban design and neoliberalism may have originated in the lack of theoretical reflections in urban design (Lefebvre, 2003), which hinders the possibility of framing critical approaches to its practice (Cuthbert, 2011). Instead, neoliberalism has been efficient in filling the theoretical hollows of urban design and dominating it.

From the theoretical articulation with neoliberalism, Alexander Cuthbert asserts that urban design provides theoretical resources for an urban political economy (2006a) because in the current scenario urban development is fundamental for absorbing value from capitalist activities. Cuthbert explains that urban design has efficiently adopted frameworks from history, philosophy, politics, culture, gender, environmental sciences, aesthetics, typologies, and praxis in its approach to explaining the conditions and complexities of urban forms in the contemporary context, which also serves to illustrate how capitalism works in cities (Cuthbert, 2011). Particularly, I am concerned about how urban design may actually become an influential discipline for pursuing social justice when it is facing the powerful forces of capitalism (Grazian, 2004; Harvey, 2009; Swyngedouw & Heynen, 2003). The lack of theory hampers the possibility of making urban design an autonomous disciplinary field, and, in the absence of a self-consciousness of urban design's own political dimensions, neoliberal domination will continue. In this case political action implies that urban designers have to recognise neoliberalism as a main contextual feature of their practices. It will require the study of the urban beyond the analysis of urban forms in order to enter the politics of space, acting more consciously with respect to their social responsibility, building a disciplinary agenda, and aiming to provide a different idea of an urban society in which the different classes of society would thrive equally. In other words, I will explore the political side of urban design. I do believe that this politicisation of urban design could follow the example of other disciplines, such as lawyers with human rights, referred to in the Universal Declaration of Human Rights; medical doctors with abortion rights, fostered by the Human Reproduction Programme of the World Health Organization; and scientists with global warming, who are supported by the Paris Agreement. While different disciplines are advancing towards a shared ethical commitment as a baseline to address their own politics, urban designers remain mostly oriented towards providing solutions, avoiding deep reflections on the processes and the modes of such production, including its social and political implications.

Some small-scale experiences aim to deliver a better-built environment by contesting neoliberalism, although broader reflection is needed to scale up their influence, passing from acupunctural practices to metropolitan applications. The scale-up is a big challenge when one considers planetary urbanisation based on a global capitalist network (Brenner, 2013, 2016; H. Lefebvre, 2003). Thus, a theory of urban design should also embrace a radical stance for contesting neoliberalism from the processes of the production of space.

Urban designers that already took a critical stance on the theory of their practice are questioning the disciplinary field and demanding more profound reflections on its scope, assuming new responsibilities with the urgencies of society. Kanishka Goonewardena (2011) establishes key questions for the urban design discipline, explicitly asserting the co-dependence of capitalism and the processes of the production of space. Furthermore, Goonewardena demands disarticulation between capitalism and urban production for developing what he called critical urbanism, a way of critical thinking in order to inform practice. Margaret Kohn proposes the use of critical theory borrowed from political science to analyse urban processes. In her interpretation, "critical theory is an approach that reads the city itself as a text in order to reveal patterns of domination, exclusion, and power relations that are difficult to recognise because of the way that they are taken for granted in our experience of daily life" (Kohn in Banerjee & Louaitou-Sideris, 2011, p. 195). Accepting the question proposed by Goonewardena as a reflective challenge, I will enquire into the theoretical problems inherent in articulations between urban design and neoliberalism.

Spatial dimensions of neoliberalism

Neoliberalism is an extremist implementation of capitalist principles wherein the means of production are privately owned and run by a capitalist class for extracting value, while workers work for a salary or wage without owning capital or the products of their own work (Zimbalist & Sherman, 1984). In neoliberalism, these assumptions expand from economics to the whole social spectrum, and the promoters of neoliberalism state that society will thrive if a free-market political economy controls it (Harvey, 2005). Thus, neoliberalism is a theory, an ideology, and a political project that emerged from an influential group of liberal thinkers (Mont Pelerin Society, 2016) who organised to contest Keynesian economic theory.[1] In a post-war context Europe and the United States manufactured an alternative governance model that worked towards the realisation of economic freedom for a social class. In practice, neoliberalism is based on the laissez-faire economics, whose goals were economic liberalisation, defence of private property rights, governmental austerity, and free trade. Neoliberalism situates the private (individual, enterprises) at the centre of the economic activities. Simon Springer explains that neoliberalism is a set of political, economic, and social arrangements that emphasise market relations, reorganise the state for facilitating business, and stress individual responsibility (a supposed merit) in achieving success in everyday life (Springer, 2016). As public policy, neoliberalism is a governmental model that eliminates control over prices, deregulates capital markets, fosters free international trade, and advances towards the privatisation of social security (Boas & Gans-Morse, 2009). David Harvey maintains that neoliberalism may be seen as an utopian project of international capitalism to restore the conditions of capital accumulation benefiting the economic elites and their capacity of controlling decision making on societies (Harvey, 2005). As such, spatial disciplines are affected by the neoliberalisation of nations. There were disciplinary transformations on urban design, urban planning and architecture developed through coercive relationships between states and the hegemonic class for privatising as many aspects of society as possible in which city making played a key role.

As Nik Theodore et al. explain, cities became central to the reproduction and continual reconstitution in recent decades in which the city has become a strategic target to neoliberal experimentation, incubating ways of reproducing neoliberal regimes (Theodore, Peck, & Brenner, 2012). While the urbanisation of neoliberalism refers to a broader scale of analysis, its representation at microscale is significant for getting into how this ideology has transformed the way we see cities. For instance, neoliberalism fosters individualism (Stiglitz, 2010), and one of the main spatial expressions of individualised space is the dwelling. Therefore, under neoliberalism, the push for home ownership by governments has been a common feature (Rossi & Vanolo, 2015). Starting in the early 1990s, after the Washington Consensus, banks reduced the requirements for credits on mortgages, which facilitated the access to home ownership through debt that involved the financial sector more actively in the production of spaces (Fernandez & Aalbers, 2016). The access of low-income communities to multiple mortgages without financial background checks and the coercion of the financial sector to maximise the earnings from the housing market were the main causes of the great financial crisis of 2008 (Fernandez & Aalbers, 2016; Follain & Giertz, 2012; Rossi & Vanolo, 2015). It is important to highlight, because the greatest crisis that neoliberalism has faced emerged from the processes of the production of spaces. Therefore, a critical theorisation of neoliberalism from the perspective of urban design seems like a suitable approach to unpacking how this ideological project affected cities. Because of the subprime crisis of 2008 and the real estate bubbles in Europe, neoliberalism entered a profound crisis of credibility, and a significant number of political manifestos began to circulate announcing its end. Neil Brenner, Nick Theodore, and Jamie Peck refer to this stage with a metaphor of a dead but dominant political-economic system that entered a zombie phase in which its brain stopped working and all the body remains were just reflex reactions to feed its incessant hunger for profit (Brenner, Peck, & Theodore, 2012).

In theorising this *walking dead* ideological project, Paul Mason developed an agenda for what he named a "post-capitalism" stage. He stated that, in order to replace neoliberalism, "we need something just as powerful and effective; not just a bright idea about how the world could work but a new, holistic model that can run itself and tangibly deliver a better outcome" (Mason, 2015, p. 13). A different future, in the view of Slavoj Zizek (2015), may be possible only with a highly bureaucratic but efficient state, and in his opinion we should not be afraid of a huge state apparatus if it ensures the emancipation of social relations. In Mason's proposal, such a new paradigm has to be based on micro-scalar economic relations inserted into a globalised network of social interactions. In line with transformative agendas, Philip Smith and Manfred Max-Neef (2011) proposed that change has to start in economics schools because that is the place where neoliberal ideology was and is reproduced. Both demand a new ethos for economics and a new teaching paradigm. Besides, Smith and Max-Neef's agenda of transformation requires more respectful social relations, human-scale development, and eco-municipalities, among other ideas. Similarly, David Harvey (2014) proposed 17 strategies for a political praxis for a post-capitalist future. There are more examples of agendas for a post-capitalist future, which represent radical approaches to the neoliberal loss of credibility. Indeed, even Joseph Stiglitz, former director of the World Bank, has exposed the necessity of overcoming neoliberalism, saying that it boosted inequality and dependency on finance and pushed the world into the crash of 2008, proving that neoliberal ideas were a failed political-economic system (Stiglitz, 2017). Neoliberalism expanded the idea of an entrepreneurial society and consequently evolved to include the idea of entrepreneurial cities as spaces where politico-economic elites work together for experimenting with innovative ways of capital accumulation which imply a series of socio-spatial restructuring processes as consequence of the speculative dynamics that have driven these entrepreneurial strategies (Rossi & Vanolo, 2015). The neoliberal regime of institutional governance that facilitated capital accumulation is a process in which the capitalist state is reshaped following the logic of entrepreneurialism in which business and political interests are aligned (Tickell & Peck, 2002). Despite the criticism, 11 years after the crisis neoliberalism remains in its "walking dead" phase, dominating the scene with its profit-oriented decisions. As Slavoj Zizek said, nowadays it seems easier to imagine the end of life on Earth than a modest transformation in the neoliberal modes of production (Zizek, 2012).

Critical reflections on the neoliberalisation of urban practices are abundant (Brenner, Madden, & Wachsmuth, 2011; Fezer, 2013; Harvey, 2012; Spencer, 2016), but the image and the design of the post-neoliberal space are an open field for exploration because not many feasible urban design projects for a post-neoliberal city have as yet been presented. What does a post-neoliberal city look like? A radical urban design has not developed this idea of the future. Hence, the post-capitalist humanity remains a fantasy without a space to inhabit. Instead, urban design projects still reflect continuity with the current neoliberal ideology. At this time, proposals for overcoming neoliberalism lack a spatial project. The post-neoliberal city does not exist as a virtual object constructed as a consequence of a series of discussions on how to overcome neoliberalism through urban production processes. Without an imagined future, it is hard to convince society to struggle to overcome neoliberalism.

Santiago de Chile and neoliberal urban development

Urbanism (*urbanisme*), in the French tradition, is a lens and a practice for analysing, theorising, and shaping cities, so when Lefebvre criticises its role in society, he refers to an intellectual approach to spaces as well as to its practices. However, in Anglophone academia, urbanism is divided into different sub-disciplines such as urban planning, urban design, urban studies, urban history, urban sociology, and even architecture. Instead, in other places, all these disciplinary scopes are grouped under a disciplinary field and study subject named urbanism (Barnett, 2011). Notably, the practitioners who shape cities are named urban designers, and they also seek to theorise the practices of the production of space. Situated in the topic of this book, I will explore whether urban design is a superstructure of a neoliberal society, whether its ethos aims to foster organisational neoliberalism, and how urban design contributes to the production, consumption, and distribution of capital by ordering the space, using Santiago de Chile's *urbanismo* as a source of data. This distinction is important because the focus of the research is the disciplinary fields involved in the design processes of Santiago de Chile, studying its urban form and the decision environment under neoliberalism. Hence, in studying this city I draw upon urban planning, urban design, urban studies, urban history and urban sociology.

Gustavo Munizaga is a Chilean urban design specialist who argues that urban design works in between social theory and architectural theory, and for the particular case of Chile it is possible to recognise a Diseño Urbano de Mercado (Market-Oriented Urban Design), focussed on supply and demand of real estate investors, and a Diseño Urbano Social (Social-Oriented Urban Design) that produces an urbanism aiming for social justice and contesting capitalism (Munizaga, 2014). Alejandro Aravena explains that discussing the future model of the city of Santiago is a waste of time. Instead, he claims, *urbanistas* should reflect on how a given city with given conditions should face its extension and densification to become a better city (Aravena in Galetovic, 2006).

From the return to democracy in 1990, Santiago as an urban phenomenon has experienced significant transformations in relation to flexibility in the urban development processes, the dominance of market-oriented policies in deciding the future of urban areas, changing lifestyles of inhabitants, increasing needs of mobility, and the new means of social communication systems. Because of the speed of these changes, the city of Santiago has become more fragmented and uneven (Bresciani in Greene, Rosas, & Valenzuela, 2011). In Chile, the primary strategy for responding to these changes is through urban projects, understood as specific urban interventions that aim to resolve or relieve certain urban issues or gaps by providing better public goods (Greene et al., 2011). However, the metropolitan form of Santiago has been shaped by policies favouring urban sprawl (Figure 2.1).

An urban development model under a neoliberal regime was implemented for the first time in the world in Santiago de Chile, by the urban policy introduced in 1979 (Lawner in Boano & Vergara-Perucich, 2017). This policy declared that the urban space should be designed and controlled by market rules (supply and demand, cost-benefit, and profitability of interventions).

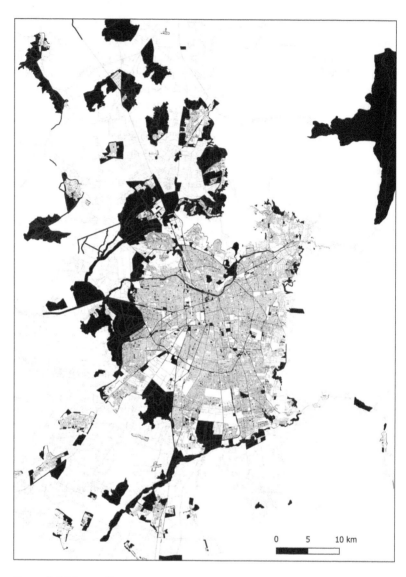

0   5   10 km

Figure 2.1 Metropolitan Regulatory Plan of Santiago number 100 (Plano Regulador Metropolitano de Santiago #100 or just PRMS100). This map represents the only legal metropolitan instrument for defining the urban form of Santiago. Basically, it defines the extension of the city. The other instruments depend on each *comuna* and the mayor's criteria. Santiago has 32 *comunas*, without a metropolitan authority for coordinating the urban transformations.

Source: Own elaboration using shapefiles available at www.minvu.cl.

After some modifications in 1985 (Gross, 1991), space was then organised as a public-private agreement between the state and free-market agents. For Andres Solimano, the neoliberal model was beyond a mere economic programme of market liberalisation because it was also an attempt to impose a new set of values in order to change the culture of Chileans, making them consumption oriented and fostering the idea of entrepreneurialism in an effective cultural revolution (Solimano, 2012). It was a utopian idea coming true by idealising free market, promoting individualism as ethos, and installing profit as goal for education, health, pensions, and infrastructures. Solimano theorises this utopia for the Chilean elite as the *neoliberal trap*, an ideological black box that framed Chilean society and forced it to remain within the limits of what capitalists defined as the priority. As discussed in Chapter 4, in 1979, the Chilean dictatorship promulgated the first neoliberal urban policy in the world, inspired by a paper by Arnold Harberger – Milton Friedman's colleague and economic advisor to Pinochet's dictatorship – in which he stated that limiting the urban area of the city by a regulatory instrument was the reason for the unbalanced differences in land value between urban and rural areas.

For Harberger, the solution to this inequality was to eliminate the regulation of urban boundaries, facilitating the sprawl of the city beyond its current limits. Specialists widely criticised the implementation of this policy from its beginning because it fostered residential segregation (Sabatini, 2000), as is shown in Figure 2.2. For Pablo Trivelli (1981), this was a failed attempt of public policy to improve the speed of urbanisation and the provision of housing for the poor. Trivelli exposed the contradictory nature of this policy. Notably, he criticised the absence of a plan for the future, an imagined reality of Santiago after the implementation of the policy. Also, he pointed out the senseless idea that land was a non-scarce resource, and he demonstrated why this policy increased the price of urban land instead of reducing it. Trivelli argued that Harberger's idea – and the urban policy of 1979 – was a fiasco. In the end, the urban policy of 1979 accelerated the formation of informal settlements and exposed the deficiency of roads, facilities, and the plans for reducing contamination in the city. As one of the results of this failed urban policy, in 1989 the number of households living in informal settlements was close to a million. The problematic urban policy, together with the earthquake of 1985, forced the government to develop a new National Policy for Urban Development that year. For Patricio Gross (1991), this new policy amended some of the mistakes of the previous one but kept the logic of an urban development guided mostly by free-market rules and a permissive set of ad hoc regulations geared towards extracting as much profit as possible from urban development processes. The *política ajustada* (adjusted policy), as it was named, declared land a scarce resource,

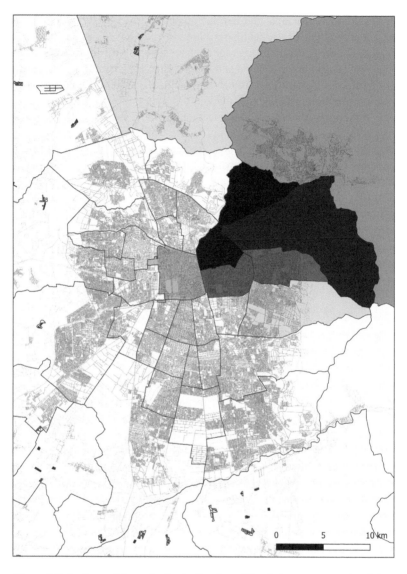

Figure 2.2 Socio-spatial typologies in Santiago. The darker areas are the richest
comunas of the city.

Source: Own elaboration based on CASEN 2015.

the market trading of which had to follow national regulations. It established the conditions for expropriating private property for infrastructural projects (although constrained by the constitutional defence of property rights) and determined that the state would define some basic regulations for the urban development market. Also, the policy defined the necessity of creating planning instruments at a municipal level, coordinated by local authorities. Finally, this policy fostered urban growth within the existing urban limits (Gross, 1991). This policy signaled the beginning of a public-private partnership in Chile, paving the way for the next 30 years of urban development. I further develop the details of this neoliberalisation of urban development in Chapter 4.

Chilean urban development has been considered exemplary for the Latin American context by economists who value prosperity and growth (Galetovic & Poduje, 2006; Glaeser & Meyer, 2002), but it presents diverse problems that for decades practitioners and scholars have striven to resolve, related primarily to the uneven development of cities and the creation of housing without ensuring the provision of public goods per area, as shown in Figure 2.3 (Rodríguez & Sugranyes, 2005).

Responding to this claim about the urban development model in Chile, in 2012 President Sebastian Piñera created a board of specialists with the remit of designing a new National Policy for Urban Development (CNDU, 2015). A wider group of actors[2] participated in this policy design than in the development of the two previous policies mentioned (Jimenez in Gobierno de Chile, 2014). In this case, the board of specialists gathered people with different perspectives about urban development, with a diverse range of political positions. Also, this policy was developed in a democracy, so its advances were open to public discussion. Even though it was born in a democratic and open manner, Leopoldo Prat (in Lopez, Jiron, Arriagada, & Eliash, 2014), the Former Dean of Architecture and Urbanism at Universidad de Chile, criticised the absence of regulations related to the human experience in the space and the low relevance given to how inhabitants value their spaces (historical, economic, political, and/or personal values).

Also, Prat revealed his concern over the absence of aspects related to the geographical differences of spaces depending on their environment and the scarce theorisation on sustainability, which in Chile is an important matter because of the extreme differences between the north, the centre, and the south. Despite this criticism, the policy advanced more substantially than the previous ones towards the production of fairer cities. The national policy of urban development emerged after the earthquake of 2010, reflecting the need for new instruments in the twenty-first century. A main problem that the policy addresses is spatial segregation, considered to be the most visible expression of how economic growth is misdistributed throughout

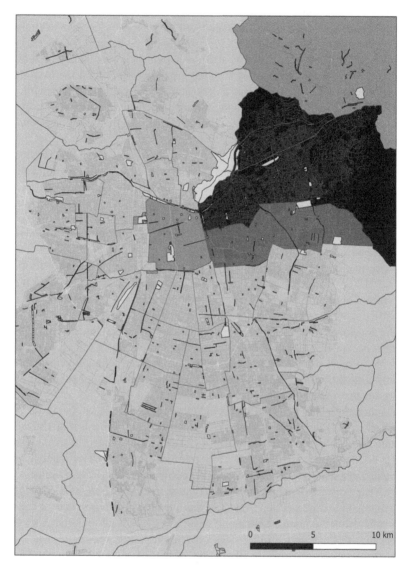

Figure 2.3 A parallel between public spaces in high- and low-income areas of the city.

the urban fabric of Santiago. Indeed, Santiago is the most segregated city among OECD members (OECD, 2013).

Usually, the cause of this segregation is attributed to public housing policy. "In order to reduce the accumulated housing shortage and to control the slums and poverty-stricken urban settlements, the Chilean government implemented large-scale housing production during the 90s. This policy was relatively successful in achieving its main goal of reducing the housing deficit from 771,935 to 543,542" (Rodríguez & Sugranyes, 2005), but it generated residential segregation as a collateral effect (Lambiri & Vargas, 2011). For Francisco Sabatini, the most complex problem of spatial segregation is the stigma attached to being poor and also to living in a neighbourhood for the poor (Sabatini, Wormald, & Rasse, 2013). On the other hand, segregation in business terminology could be understood as a strategy for allocating resources, better known as segmentation.

This means that the market has to be divided into clusters of clients depending on their demands, purchasing capacity, and interests (Silbiger, 2009). In the neoliberal city, the design of the spaces may well fit into the same logic, creating spaces for the poor and the rich and dividing qualities and supply depending on the customer (Figure 2.3). In doing so, urban designers are the executors of these urban policies. Therefore, their participation in developing these regulations not only is significant but should help to inform the creation of a theoretical position on the production of spaces under a neoliberal regime. In the whole discussion of the new urban policy, no images or ideas of future cities were presented.

Gentrification is another consequence of urban design under neoliberalism. It is related to the segregation of space, as both follow the logic of segmenting the city by clusters of consumption. Ernesto Lopez-Morales defines four features of gentrification in Santiago: state subsidies for the demand for housing for upper-income classes of society, which facilitates the expulsion of lower-income communities; a disparity in capturing rent between the original owners and the large-scale redevelopers; the purchasing power of developers when acquiring land; and the zoning method for defining land use and hence localising rent gaps in the urban fabric. These processes contribute to changing the scale of segregation in the whole city, pushing people to more affordable areas (Lopez-Morales, 2016). While in the case of segregation the cause seems to be public policies, in the case of gentrification the role of developers as urban speculators appears to be a significant factor, although the two processes are deeply related.

In this research, both segregation and gentrification are symptoms of the urban design illness. Looking for its cure, Ivo Gasic Klett (2016) and Rodrigo Cattaneo Pineda (2011) have pointed out the infection, the possible main cause of the zombie phase of urban design: the financial system. The role of the financial institutions in funding urban development projects is directly related to the verticalisation of Santiago's skyline and to the localisation of urban projects in specific areas of the city. Cattaneo states that the financialisation of Santiago determined specific typologies of space, selecting those that were more profitable for the system. Following this interpretation, it is possible to assert that the market has defined the new urban form of Santiago and that urban design practice is framed entirely by the interests of the financial institutions, either private (banks) or public (state). The financialisation of urbanism is a symptom of a so far incurable illness named neoliberalism.

Beneath the neoliberal ideology, a model of spatial production had advanced and refashioned itself not only for using urban development for the sake of capital but also as a means to relieve the social claims for solutions in relation to urban life. Neoliberalism has been adapted, and instead of refusing the demands for better provision of public goods in cities, it is using this claim for generating new resources emerging from the fiscal budget in the form of subsidies or tax benefits (Dattwyler, Christian Voltaire, & Santana Rivas, 2017; López-Morales, 2009). A neoliberalism with a human face is a sort of privatisation of social demands defining paths for both legitimate claims emerging from society and validation of capital goals as part of the solution. As Rodrigo Hidalgo Dattwyler et al. (2017) observed, social demands emerging from critical theory such as spatial justice or the right to the city may be transformed into new resources for the sake of the neoliberal hegemony.

Recalling the definitions of Cuthbert (2006) and Carmona (2014a) in their acknowledgement of a gap in knowledge in urban design theory, the influence of the financial system, the capitalist state, and the profit-oriented goals of design are fundamental to understanding the nature of urban design under neoliberalism.

Concluding remarks

The aim of this chapter was to introduce the relevant literature that informs the theoretical position taken in relation to the disciplinary field of urban design and its potential connections with neoliberalism. Also, this chapter introduced briefly how these two components are present in the context of Santiago. The current discussions about the urban effects of neoliberalism are not particularly precise when analysing how this ideology transformed the way of designing the space between buildings. The focus of the studies on urban design under neoliberalism is on the space as a resource for capitalists, but not much has developed in reference to the decision environment in which urban designers work. Also, reflections about the ethos of urban designers when defining the urban form in the case of Santiago are not abundant. There is a gap in the literature in critiques of the role of urban designers in shaping the city of Santiago. The decision environment of urban designers in the case of Santiago remains obscured. Investigating it may illuminate the way neoliberalism actually influenced the decision making that allowed the possibility of an actually existing urban design under neoliberalism. I argue that decisions made by urban designers under neoliberalism are the main engine of the spatial reproduction of neoliberalism, which are reflected in Santiago's urban life.

While urban design and neoliberalism are categories for analysing a phenomenon, in the end, people are behind these activities. Neoliberalism is not a ghost or an invisible hand that moves automatically. Neoliberalism is the conceptualisation of a political-economic project related to free-market economics and monetarist theories for shaping the social relations that have been implemented by individuals with certain interests and beliefs. This research contributes by discussing the role played by urban designers – as a disciplinary body – in the reproduction of neoliberalism, with particular focus on the ethos and contradictions faced by these practitioners under such regime.

Notes

1 The Keynesian economic theory was developed by the British economist John Maynard Keynes. This theory argued that the markets are not self-regulated and that the state has to play a key role in the economic development of countries. For instance, Keynes stated that full employment is vital, as is ensuring social security, so that workers can spend their salaries on goods and services.
2 The process took three years until its final text was published, and scholars, practitioners, social organisations, unions, politicians, and entrepreneurs were invited to take part in its formulation.

3 Spatial dialectics of Santiago's urban land

Introduction

This chapter is focussed on the historical dimension of Santiago's urban space from the beginnings of the twentieth century until the end of Allende's government just before the implementation of neoliberalism. The objective is to uncover how the urban space of Santiago has been shaped by a dominant class mostly focussed on creating wealth and defending property rights. This chapter presents neoliberalism as a political-economic project which emerged in the twentieth century, but its principles and ideas were already present in the mindset of its oligarchy. In order to facilitate the narrative, when the city of Santiago is mentioned in the text, it means that I am writing about the Gran Santiago or the Great Area of Santiago. The concept of Gran Santiago was introduced in 1960 by Justo Pastor Correa and Juan Honold, and it was a category of geographic scale for the city in the Metropolitan Master Plan of 1960 (Gurovich, 2000) developed by these *urbanistas*. However, for 400 years the city of Santiago has been better known in the literature as just Santiago.

The urban history of Santiago may be analysed as a spatial dialectic, which means that the analysis of its historical development situates land at its centre as the desired object disputed among members of society. Struggles over control of the land have been the engine of Santiago's history. In the case of Santiago, the historical narrative articulating urban design and the interests of the dominant class through spatial dialectics have not been developed before. Indeed, this chapter may serve as the starting point for a further deeper historical analysis of the relationship between space and capital in this city. It is relevant knowledge that appears scattered in the literature but that has not yet been completely organised systematically. In building this chapter, I have used the work of Luis Vitale, Armando de Ramón, Gabriel Salazar, and Patricio Gross. Also, I employed historical economic data developed by the Department of Economics at the

Universidad Católica de Chile. Various archives were consulted for making this chapter accurate but not overwhelming. I wanted primarily to make the case for Santiago as a city built for capitalist interests and understood as an asset for extracting value from spatial transformations.

This chapter presents first the period characterised by the imposition of private property by the oligarchs and by a specific mode of shaping the city, based on the aggregation of plots of privately-owned land in which the dispossessed were seen as a problem to subjugate. A second period started with the emergence of urban design (*urbanismo*) as a means to divide the city between the *barbarians* (urban poor) and the *civilised* (oligarchy); a final period, called *the city of the masses*, was characterised by the state's strong role in planning the city, in which the whole society was organised to alleviate urban poverty. The city of the masses ended in 1973 with the coup d'état led by Augusto Pinochet and with the subsequent implementation of neoliberalism in 1975. When planning this chapter, I had to exclude some key themes and topics[1] of Santiago's history in order to reduce the length and also to synthesise the events that, from my perspective, contribute best to understanding the entangled relationship between the disciplinary field of urban design and the political economics that have shaped the city.

The *conventillo* and the social question: the city as a political arena of spatial contestations

In the beginnings of the twentieth century, the technological advances in agro-industry implemented in California, Australia, and Great Britain undermined significantly the demand for grains from Chile. The contribution of agricultural activities to the GDP reduced from 12% in 1879 to 6% in 1912 (Díaz, Lüders, & Wagner, 2016). This agricultural crisis was accompanied by a decline in the production of saltpetre, a principal economic resource at that time and demand for which dropped by 58% between 1916 and 1924 (Gonzalez Miranda, 2014). Both activities represented important job providers in Chile, and both were based on exploitation of the countryside, either in agricultural fields or in mines. These economic changes may explain the huge increase in migration from the rural area to the urban centres after 1900. Historical research reveals that 960,298 people moved to Santiago between 1907 and 1960 (De Ramón, 2007). This number is significant considering that in 1895 the population of Santiago was 256,403 and by 1960 it was 1,521,831, which represents an increase of 493%.

This intense process of migration from rural to urban areas demanded significant transformations in the urban space. While in 1850 the rural population represented 80% of the total population of Chile, in 1895 it had fallen to 54% (Díaz et al., 2016). Between 1890 and 1940 the number of people living in urban areas increased from 1,223,407 to 2,639,311; Santiago's population increased from 290,000 in 1895 to 952,075 in 1940, which represented a concentration of 36% of the whole urban population. Despite the economic growth, the political organisation, and the advances in the formation of a democratic republic, Santiago was not prepared for such challenges, and new knowledge was needed to reconcile this demand for space with the needs of its inhabitants. After years when the city development focussed on private spaces, public space became as an urgent matter. Dispossessed people coming from rural areas were forced to live in precarious and unhealthy conditions, which was damaging to how the elite perceived their city. Nevertheless, many landlords were taking advantage of this migration and developing a series of businesses related to the provision of cheap accommodation in Santiago, renting rooms to people or families looking for opportunities in the main city of the country. This new scenario reshaped the city, and new typologies emerged to cover the demand for shelter.

Among them, the most common typology of settlement occupied by the low-income inhabitants of Santiago was known as *conventillos* (De Ramón, 2007; De Ramón & Gross, 1985). It was a type of basic accommodation composed of a room without windows, toilet, or sewerage. These facilities sheltered several individuals per room, living together in a shared house, paying rent to the owner of the building, who was usually an oligarch who had built a cheap house to establish a *conventillo* business – renting low-quality accommodation at high prices. Besides the alienating nature of this business, there were the health problems associated with *conventillos*. Indeed, sanitation was a massive issue in the Santiago of the early twentieth century.

These *conventillo*s were a highly profitable business for the landlords (Gross, 1990, p. 75) because the price of renting was much higher than the investment, and most of the rented houses did not have any kind of maintenance or facilities at all. Indeed, sometimes the landlord rented only a plot of land where the tenants could build their own spaces with cheap materials (usually wood and metallic sheet). This mode of wealth creation exposed a condition that several foreign scholars found concerning. Horace Rumbolt (in De Ramón, 2007) said that this phenomenon was easily explainable because Santiago was a city created by an oligarchical government whose interest was not the public domain and collectivity but the development of private wealth. The centrality represented by Santiago absorbed most of the country's wealth, making it an expensive city, especially for the poor. For Rumbolt, it was also dramatic that the buildings of the oligarchy were built so close to slums, making misery a postcard of the very centre of Santiago

(Rumboldt, 1877 in Godoy Urzúa, 1981). Moreover, inadequate housing (either *conventillo*s or informal settlements of the homeless) constituted 70% of the accommodation in Santiago in 1870 (De Ramón, 2007).

In the nineteenth century, the urban form of the city followed private property expansion, rather than a coordinated idea of urbanity. When the number of people living in informal and precarious conditions increased drastically at the beginning of the twentieth century, it became a threat to the oligarchical idea of the city as well as to the health of the elite. The clash between two ways of occupying the city, the palace and the *conventillo*, forced the introduction of new understandings and specialisations of the urban development process.

In 1891, Chile entered into what has been named the parliamentary regime (Salazar, 2003a). After the crisis of 1875, the oligarchy organised itself to occupy positions in the state, causing a significant increase in the expenses of the fiscal administration (from 3.6% share of the GDP in 1860 to 7.6% in 1890). In the state (government, congress, and judiciary power) they found a means to preserve their status of social dominance. Indeed, the elite professionalised politics as a career (Salazar, 2015), and many of them left the life of a merchant to become mayors of a city, members of parliament, or employees of the government.

This period marked the formation of what Gabriel Salazar called the Chilean political class (Salazar, 2015); most of its members moved to live in Santiago, concentrating in its hands control of land, politics, and the economy. Charles Wiener in 1888 described the Chilean elite as a group of aristocratic oligarchs who excluded the majority from decision making on public affairs. Data confirm Wiener's perception (Table 3.1), as the incomes of the richest 10% increased from 49% of all income in 1865 to 60% in

Table 3.1 Income distribution per population income deciles in Chile between 1865 and 1930.

Deciles	1865	1885	1907	1930
1	4%	3%	3%	2%
2	4%	3%	4%	2%
3	4%	5%	4%	2%
4	5%	5%	5%	3%
5	5%	6%	5%	3%
6	5%	6%	5%	5%
7	7%	8%	6%	5%
8	7%	8%	10%	7%
9	10%	11%	14%	11%
10	49%	45%	44%	60%

Source: Rodríguez Weber (2009).

1930. Javier Rodríguez Weber (2009) estimates that in 1865 the richest 1% received 21% of income, which increased to 27% in 1930.

In shaping the parliamentary republic, the oligarchy aimed to ensure its dominance over the population and the fate of the country. Control had obsessed the Chilean oligarchs since the early years of the colony. Several historians conceptualise how the Chilean oligarchy always felt afraid of the different, of the *rotos*. For Salazar (2003a) and De Ramón (2007), this may be explained by the historical trauma of the Mapuche invasions and the destructions of cities that occurred in the Arauco War. De Ramón (2007) says that this was a constant concern in the psyche of generations of oligarchs. If, initially, the fear was of the Mapuche, it then became fear of the *roto*, and in the twentieth century the urban poor were feared. Controlling the country was a measure for controlling their fears and keeping possible insurgencies of the urban poor in line. According to my interpretation, this fear may also explain the obsessive aspiration for private property and the sacrosanct defence of the rule of law. The palace, then, was not only an asset or a space for the oligarchs to live in but also a means of isolating themselves from the scary public space. The elite used all the power of the state to defend its agoraphobic lifestyle.

The parliamentary republic produced a series of significant spatial transformations in the country, because a great share of the fiscal budget was invested in infrastructure and urbanisation (Braun, Braun, Briones, & Díaz, 2000; De Ramón, 2007). For example, the state started to buy plots of land in cities like Santiago. Indeed, given that most of the land was already in the hands of oligarchs (now politicians and governmental officers), urbanisation and infrastructural construction were a circular business in which the state financed the wealth creation of an oligarchy in crisis. Hence, the public funds were used for financing government facilities, public buildings, parks, and other resources that the city required for functioning properly. It is important to mention that the highest productivity per worker between 1890 and 1930 was in the mining sector ($10,479,774 Chilean pesos of contribution to the gross domestic product [GDP] per worker) and in construction ($5,491,733 Chilean pesos of contribution to the GDP per worker). Although manufacturing was increasing its contribution to the wealth of the nation, constructors and miners were the most contributive workers. Chilean capitalism depended significantly on producing value from land, either as minerals or as buildings.

With the migration from rural to urban areas, low-income workers changed their minds. The city of Santiago allowed people to meet in the space, and the poor realised that their living conditions were not affecting just a few but many others. The *bajo pueblo* became aware of their alienated condition. Furthermore, they had access to information, education, and books, and leaders started to emerge from the dispossessed masses. The city of Santiago was framing these realities: the urban poor and the wealthy oligarch, the *conventillo* and the palace, the civilised and the barbarian. The clash between those who owned the land and those who worked the land would produce new politics of space and an incipient contested city that would unveil the necessity of a disciplinary approach to the urban problems emerging.

It is possible to say that between 1890 and 1930 two streams of urban transformations occurred in Santiago: one related to architecture and the other related to relieving several urban problems associated with sanitation and the consequences of the explosive migration from the rural areas. In relation to architectural transformations, these received a boost with the celebration of 100 years of the republic in 1910. Significant efforts were made to build spaces that represented how this young country had thrived in so short a time. Thus, Santiago experienced the rising of the Justice Palace, the Beaux Arts Palace, the National Library, the Mapocho Train Station, the Forestal Park, the Centenario Park, and the landscape improvements on San Cristobal hill. All these projects aimed to foster a public life in the city, incorporating new functions and enriching urban life. On the other hand, while the oligarchy built these works for the sake of the beauty of the city based on an European nostalgia (De Ramón, 2007), the rest of the population started to organise themselves to demand better spatial conditions. They would find an echo in some of the politicians who were most committed to their role as representatives of the people.

One of the main concerns was related to the urgent necessity for sanitation in Santiago. This was the hygienist urban transformation, based on various regulations and analyses of the city, exposing the differences between the spaces of the poor and the spaces of the elite (Ibarra in Perez, 2016). Before the hygienist reforms, the child mortality rate had peaked at a high rate of 30% (Díaz et al., 2016). Among several causes, the urban space and the housing were considered to be the main causes of the problem, and so they were the focus of the search for solutions. The demand for shelter was much greater than the available housing accommodation considered to be healthy. In 1906 the state promulgated the Law on Labour Housing, which aimed to hygienise, normalise, and define the way to build houses. In spite of the apparent good intentions behind this norm, the state built only 396 houses between 1906 and 1925 (Gross, 1990, p. 79). Faced with this inadequate solution, the middle classes started to develop a class consciousness, while the urban poor began to organise themselves to protest against the abuses of the oligarchy, radicalising their strategies for protesting (Espinoza, 1988a).

With regard to the urban poor, the politicians considered them as a hygiene problem until 1925. Nevertheless, the problems of the lower classes were also ideological, as they started to become aware of the unjust society around them and the fact that in a democracy, they had the right to demand fairer treatment. From diverse sources and international literature, the urban poor started to look for a way to politicise their claims, exacting transformations in the political system. Capitalism was not a panacea, and the people started to conceptualise their discontent. Various members of the elite with social consciousness became critical of their privileges and supported the political formation of the lower classes.

The differences between the oligarchy and the urban poor were no longer visible only in the spatial dimension; from 1920 onwards, this issue turned into a political dimension for transforming what was considered to be injustice. The city joined both the political and the spatial to motivate the organisation of people towards a particular goal. A good example in Santiago was the Liga de Arrendatarios (Renter's League). In 1922, the inhabitants of 300 *conventillos* in Santiago initiated a rent strike because the fee for renting was sometimes higher than the total amount of their incomes – and this for unhealthy rooms, without a toilet, water, or even windows. Several strikes were organised between 1922 and 1925, when a significant event occurred. On 13 February 1925, more than 30,000 tenants did not pay their rent, and they stipulated that they would not pay their rent anymore until the fees were adjusted to a more reasonable and fair amount (Espinoza, 1988).

Veronica Salas (2000) indicates that the strike lasted six months, although Vicente Espinoza (1988) points out that it was not continuous. In the end, the movement was successful and forced the government to create a specific law. This regulation would reduce housing rents by 50%. Also, it was established that houses must be built with the minimum conditions required for living properly: they would have a toilet, windows, and a door (Vitale, 2011). This strike made clear the power of collective engagement in the fight for dignity. Consequently, new political structures were consolidated (the Communist and the Socialist Parties, for instance), and the political scene, previously formed exclusively by the oligarchy, was diversified, and the democracy became a more complex arena for organising the nation. Just in 1925, after the government of Arturo Alessandri, some legal enhancements were made for the working classes. Nevertheless, the city and its complexities required new methods to tackle its issues, and new knowledge would be imported (again) to succeed in this task. In this period, Santiago attracted people from the rural areas, and a new mode of capitalist development based on manufacturing, services, and construction emerged.

Scientific urbanism and the masses: private property at stake and the neoliberal revolution of the outraged oligarchy

Between 1894 and 1925, six plans for the transformation of Santiago were presented, but none of them was implemented. In spite of the intentions to plan a new urban form, the city continued expanding, but the plans exposed the preoccupation of politicians with rethinking the urban space. Although private property remained an elementary right, the mindset of politicians incorporated new principles to face the challenges of the migration from the rural to the urban. For instance, an interesting change appeared in the Constitution of 1925, where the expropriation of land became legal when a public project required it.

> The property right is subject to limitations or rules that demand in the maintenance and progress of social order, and, in this sense, the law could impose obligations or easements of public utility in favor of the general interests of the state, the health of the citizens and public health.
> Constitución de la República de Chile (1925, Art. 10–10)[2]

This was the only time in Chilean history that a constitutional act allowed expropriation for the sake of the common good. This modification implied that the new generation of oligarchs (those who were born and bred politicians, not entrepreneurs) had a different understanding of the public and their role in society, and the urban space would reflect this change. With an oligarchy more concerned with public affairs, the city moved to the forefront of the political dispute. The Chilean oligarchy had a profound attachment to institutionalism and an increasing devotion to politics (Gross, 1990); thus, the new urban form emerged with a set of institutional instruments for organising urban development in public works, housing, and sanitation.

Furthermore, the implementation of urbanism – as a coordinated and hegemonic discipline to guide urban development – appeared as a necessity for institutionalising the design of cities. When the elite became aware of the importance of shaping the urban setting, a mentor was needed to launch the discipline of urbanism in Chile. Rodulfo Oyarzún Philippi proposed Karl Brunner[3] as a mentor, and in 1931 he created the Institute of Urbanism at the Universidad de Chile, with a programme of Urbanismo Científico (scientific urbanism). This course served to organise theoretically the urban problems in Santiago and to develop strategies for resolving these issues. Accordingly, this programme produced the first generation of Chilean urbanists.

Urbanising and designing a proper urban space also had an institutional form represented in the Planning Department of the Public Works Ministry, inaugurated in 1934. The institutional approach to urban affairs was a response of a more democratic republic to the migration from rural areas to the city. On the other hand, for the Chilean elite, democracy seemed profitable, thereby taking care of the voters' wellbeing, so maintaining an illusion of social justice was a priority in order to preserve the political and economic power in the hands of the oligarchy.

The improvement of urban spaces was a visible and concrete public policy that required the investment of public funds to maintain the illusion of democratic society. However, in the end, the country remained in the hands of the oligarchy. For Gabriel Salazar, between 1932 and 1957 the country was ruled by a populist oligarchy centred on using the state as resource for maintaining a dominant position in society aiming to provide relief to the social question and also control the national budget. Indeed, Salazar showed that 60% of the Chilean political class between 1932 and 1957 was composed of politicians and businessmen (Salazar, 2015, p. 987).

Building the city created urban products (capital) and employment. Mainly, urban development was a great business, besides producing enhancements in the everyday lives of the voters. The urban space would then become a platform for both creating wealth and fostering democracy.

Several public policies aimed to alleviate the issues related to the urbanisation of the population from the country. In March 1925 (Law 308 of 1925) the Superior Council of Social Welfare was created. Employing funds from the Mortgage Credit Bank, the Council built 43 towns. In February 1931 (by order of Law 4931) the Popular Housing Bureau replaced the Superior Council of Social Welfare in order to organise the irregular settlements in cities. Nevertheless, the state was not capable of covering all the demand for housing. In 1938, Law 6172 allowed the Popular Housing Bank to build housing with public insurance funds.

All of these efforts were not enough to resolve the ongoing urban problems. In this context, scientific urbanism provided a pathway for organising the city based on facts, ideas, and resources. In 1938, during the first national conference of Urbanism in Valparaíso, the audience agreed to prioritise the implementation of urbanism as a science and art in order to organise collective life (Munizaga, 1980). Earlier, in 1934, Karl Brunner outlined the proposal for Santiago's urban planning, and in 1939, Roberto Humeres created the Official Plan of Urbanisation for Santiago (POU) in order to implement the ideas proposed by Brunner in 1934.

The plan was approved, but the authorities ignored many of its ideas. While urbanism as a discipline aroused interest among scholars and specialists, the oligarchy kept its distance (Gross, 1991). It is interesting to discuss some of the ideas included in these plans. One of the main proposals was to zone the central areas of Santiago in order to construct buildings of similar heights (eight floors) and to maximise the subdivisions and plot allowances. These new neighbourhoods considered eight-storey buildings for housing and one plaza for every five blocks. This central district aimed to have characteristics similar to those of Paris's and Vienna's central areas. Also, Brunner's plan proposed the creation of working-class towns near the central district of the city. Likewise, housing typologies were proposed for some areas that were similar to garden-city models and targeted for higher classes. Most of these ideas were based on the idealised vision that Brunner had of Vienna: architecture composed of medium-height buildings, wide avenues as boulevards, and valorisation of the design in order to create something beautiful (Gross, 1991, p. 35). Some of the basics of this plan lasted for decades as an aspiration for Santiago.

The idea of Brunner was profitable for the state and investors, but not overly so. It is important to mention that the Chilean elite wanted to restore the profitability of its activities to the way it was in the nineteenth century. Urbanisation was an efficient path towards recovering great revenues from productive activities in the city. Some strategies of Brunner's plan succeeded, but in the end the desire for profitability prevailed. For the oligarchy, wealth creation was more important than creating a cultural expression of society in cities, as Brunner proposed. Thus, the scientific urbanism – as a discipline – was useful for justifying the short-term investments of the fiscal budget in a certain type of project, but none of the long-term urban plans was ever implemented. The very nature and main goal of this scientific urbanism was the long-term planning of urban life (Brunner, 1939), and that is precisely the realm where it failed most visibly – its efficiency for the improvement of the wellbeing of the whole society was questionable.

Instead, it is possible to say that the oligarchy instrumentalised scientific urbanism to justify the investment of public funds in short-term projects that ensured the creation of wealth and fostered the illusion of social progress for the voters. For instance, between 1930 and 1973 scientific urbanism served to scale up an efficient method for defending private property through public policies that also segregated the rich from the poor, using pseudo-scientific arguments. Furthermore, this discipline transformed urban development into a highly efficient method for producing wealth in the form of spaces (Table 3.2), representing an average 45% of GDP. Once scientific urbanism was validated, the organisation of urban life followed the rule of economic efficiency. Measurements and statistics confirmed the primary design inputs.

The presence of statistics to justify spatial transformations is striking when reviewing journals of architecture and urbanism between 1940 and 1975, mixing spaces with social measurements and costs.[4] The statistics, then, subjugated design, and beauty became a value hard to measure, so

Table 3.2 Urbanisation as percentage of GDP in Chile.

Year	GDP in urbanisation (construction+utilities+ transport and communication+ housing)	Real GDP (2003)	Share of total GDP
1940	1,979,191	4,894,316	40%
1941	1,984,780	4,902,180	40%
1942	2,001,462	5,063,603	40%
1943	2,051,304	5,208,526	39%
1944	2,191,744	5,307,108	41%
1945	2,374,155	5,765,048	41%
1946	2,679,086	6,258,938	43%
1947	2,437,525	5,585,720	44%
1948	2,494,383	6,512,765	38%
1949	2,345,716	6,372,124	37%
1950	2,645,516	6,686,125	40%
1951	2,734,635	6,977,024	39%
1952	2,840,754	7,426,609	38%
1953	3,123,655	7,986,501	39%
1954	3,168,659	7,727,830	41%
1955	3,113,433	8,018,659	39%
1956	3,292,636	8,153,261	40%
1957	3,579,451	8,989,939	40%
1958	3,836,435	9,483,619	40%
1959	3,597,174	8,946,687	40%
1960	4,024,639	9,688,927	42%
1961	4,037,527	10,152,296	40%
1962	4,321,021	10,633,377	41%
1963	4,940,239	11,306,061	44%
1964	4,789,879	11,557,631	41%
1965	4,840,924	11,651,051	42%
1966	5,176,353	12,950,278	40%
1967	5,410,172	13,370,719	40%
1968	5,679,881	13,849,417	41%
1969	6,171,759	14,364,727	43%
1970	6,471,747	14,660,107	44%
1971	6,984,726	15,972,954	44%
1972	6,598,174	15,779,175	42%
1973	6,356,587	14,901,061	43%
1974	6,644,396	15,046,274	44%
1975	6,012,388	13,103,758	46%
1976	6,065,857	13,564,778	45%
1977	6,606,376	14,902,086	44%
1978	7,329,564	16,126,649	45%
1979	8,312,631	17,462,203	48%
1980	9,482,098	18,849,559	50%
1981	10,359,067	20,020,532	52%
1982	9,411,134	17,300,151	54%

1983	8,008,821	16,815,394	48%
1984	8,042,066	17,805,071	45%
1985	8,892,094	18,155,525	49%
1986	9,435,271	19,171,550	49%
1987	10,115,638	20,412,276	50%
1988	10,770,941	21,911,017	49%
1989	11,841,380	24,228,285	49%
1990	12,324,922	25,142,427	49%
1991	13,388,425	27,136,661	49%
1992	15,148,659	30,438,172	50%
1993	16,447,125	32,559,288	51%
1994	17,217,016	34,416,719	50%
1995	19,004,777	38,028,587	50%
1996	20,349,408	40,831,593	50%
1997	21,844,223	43,526,542	50%
1998	22,778,351	44,944,336	51%
1999	22,235,188	44,616,344	50%
2000	23,268,649	46,605,195	50%
2001	24,205,130	48,165,621	50%
2002	24,967,429	49,209,326	51%
2003	25,911,045	51,156,415	51%
2004	27,486,205	54,246,819	51%

Source: Own elaboration.

it disappeared from the decision-making processes. Private construction companies built most of the projects in Santiago. In order to be suitable for working with public funds, these companies needed to demonstrate annual profits of more than 50,000 pesos when applying. This criterion segregated the companies able to bid. In the long run, this rule hindered small companies and tended to benefit those with greater capital.

The investment of the public budget in building facilities and public goods to benefit the poorest groups of the society paradoxically increased the wealth of the bigger companies. At the beginning of this period, the profitability of construction activities was regulated by each public institution. Law 8,412 of 193,834 clearly defined that the state was responsible for regulating the use of public funds by building companies, determining a maximum allowable gain. However, these regulations faded with time. More liberal policies were implemented to increase the profitability of construction, while the lobby for these activities was organised. The construction business was full of opportunities provided by several reconstruction processes after earthquakes, demographic changes, and the progress in construction technologies. Aware of these advantages, in 1948 different building companies created the Chilean Chamber of Builders (CChC) in order to create a political force capable of influencing the authorities for the sake of their business. The CChC today represents one of the main economic actors in the country, and its main goal is to defend private property. The spatial expressions of these approaches matched the paradigmatic approach to urbanism promoted by the CIAM. The CIAM was an international organisation of architect that had a manifesto redacted by Le Corbusier as main figure and leader. This manifesto fostered the worldwide adoption of the architectural principles of the Modern Movement, seeing the political and economic value of architecture to improve urban life by a universal approach to urban planning based on economic efficiency, massive production of housing and central control over city making.

Le Corbusier transparently exposed that urban planning is a way to make money (Le Corbusier, 1967), an interpretation of urbanism suitable for the mindset of Chilean oligarchs. Indeed, the whole modern movement fit with the regular requests of the oligarchy: highly profitable activities (urbanisation), elite-controlled decisions (specialists), democratic concepts (a good city for everyone), and application of trends that were popular in Europe. Urbanists defined the use of land for buildings, agriculture, mining, and transport. A body of experts designed and implemented public policies, composing then a technocratic body supposedly free of ideology and independent from political affairs for shaping the city (Gross, 1991, p. 36).

These urban experts defined a series of strategies for dealing with the demand for jobs in cities and the organisation of the urban expansion. For example, Law 7600 of 1943 determined that companies must invest 5% of their profits into housing for their workers. Another good example is Law 9135 of 1948, which created a series of benefits to facilitate homeownership for the middle class, fulfilling one of its demands (Gross, 1991). In 1952, President Gabriel Gonzalez Videla created the Housing Corporation (CORVI), whose objective was building housing and public spaces, with a special emphasis on low-income communities. In order to advance with a legal framework for the city, in 1953 the General Law of Construction and Urbanization was updated to make it suitable for the changes that had occurred in urban areas. In this update of the law, it was determined that there was a need for the urgent creation of municipal master plans for defining the function of urban areas. Through these regulations, the state attempted to deal with housing solutions and city planning in conditions of economic scarcity. These new institutional regulations were the result of the implementation of scientific urbanism in Chile, although it never crossed the frontiers of the interests of the hegemonic class. Moreover, the demand for housing grew faster than the capacity of the state to provide solutions. Society organised unions and community organisations to build an agenda capable of empowering its voices and generating changes that would transform the face of the city.

One of the better examples of this emerging social empowerment refers to people producing their own housing. In 1957 the illegal occupation of lands in La Victoria, near Santiago's downtown, began. Thousands of families constructed their houses in a self-managed approach. Various political institutions and groups such as the Communist Party, the Socialist Party, students and workers unions, and the Catholic Church assisted people with organising these *tomas*.[5] These occupations contested a sacred right for the oligarchy: private property. La Victoria soon became a referential experience for many other communities, despite its illegality. People went beyond the law, fighting for their right to shelter, producing a clash between their needs and the interests of capital. This happened in the very centre of Santiago. Private property was under attack. In the eyes of the oligarchy, the fear of the Mapuches, the *rotos*, and the squatters were all the same: a threat to the sacrosanct private property right.

Besides these issues related to contested territories, the economy was beginning a period of inflation that increased for decades without stopping (Salazar, 2003a). Looking for solutions, in 1958 people elected the capitalist Jorge Alessandri as president to foster productivity. While economic growth improved under his government, the wealth never reached the majority. *Tomas* continued emerging during his government, so Alessandri doubled the construction of social housing and in 1960 implemented a decree which included taxing benefits for middle-class housing (DFL2) and promoting a garden-city model of urban development, mostly located in the borders of the city. However, this policy was regressive. The high-income families took advantage of this decree and built their houses far from the city centre, using a regulation that was supposed to help middle-income families. In the interpretation of Patricio Gross (1991), the regulation of DFL2 missed a point because it "didn't restrict some areas of the city, which allowed that high-income people to exploit these benefits, settling on the east side of the city, increasing the process of spatial segregation" (Gross, 1991, p. 39). Consequently, the city stressed the localisation of people in certain areas depending on their purchasing capacity, so the richest people built their houses in Las Condes and Vitacura (east-north) and the middle classes in Providencia, Ñuñoa (centre-east), and the east side of Santiago, while low-income families occupied the rest of the city. The distribution of the richest families, middle-income families, and the lower class outlined a segregated space, which constituted one of the first symptoms that Santiago was entering a critical phase. The explosion of the urban space started with some fragments such as segregation, squatter spaces, contested urbanisms and the urban poor.

From the welfare state to the neoliberal revolution

In an attempt to control the contestation of private property, Alessandri's government implemented a modest agrarian reform to redistribute land in rural areas and to give peasants landownership rights. The agrarian reform was promoted by the Catholic Church in Latin America. Nevertheless, the liberal and market-led mindset of Alessandri's administration meant that the reforms did not alleviate the processes of contestation of private property. Looking for alternatives, voters elected Eduardo Frei, who proposed a *revolución en Libertad* (revolution with freedom) supported by the Catholic Church and the Democratic Christian Party. This revolution with freedom was a mix of socialist strategies and free-market economics, aspiring to develop a Chilean version of the Keynesian model of the state. Hence, in 1964 – when Frei took office – the state took on a redistributive role and subjugated private property rights to society's common good. Land acquired a social role.

In spite of the economic inflation, the political discourse promised a future with social justice because Chileans deserved it. People engaged with these ideas through demonstrations in the streets to transform the way that democracy was conceived. The administrations of Eduardo Frei (1964–1970) and Salvador Allende (1970–1973) embraced this challenge with political projects that represented the popular demand for better democracy. In Frei's administration the Catholic ideology of social doctrine was applied, aiming to create a more just society, using the state as an effective agent for distributing opportunities and benefits for people. The administration's strategy considered nationalising natural resources, fostering economic productivity, modernising the administrative apparatus, and developing a state capable of providing wellbeing through social services, thus covering the basic demands of human life.

Supported by the Catholic Church, Frei's agrarian reforms went beyond Alessandri's, giving land rights to peasants in order to increase the productivity of lands that in many cases were under-exploited by landlords. If peasants were involved in labouring the soil and also in the earnings, the productivity of those under-exploited plots would soar. Also, this was a straightforward form of distributing the benefits of agrarian and livestock activities among low-income communities. Indeed, the agrarian reform constituted an approach to spatial justice that also had a version in urban development. Frei launched a series of programmes for social housing in which participatory processes were included. These strategies aimed to alleviate the possible urban collapse of Santiago, given the incessant migration to the central city of the country. Beyond this agrarian reform, the government of Frei was not particularly a threat to private property, although this right was no longer untouchable. While low-income communities were contesting private ownership in the streets by occupying private plots of land, the oligarchy started to engage politically in defending its ownership rights.

In the 1960s, Chilean politics became radical. Land became an object in the political dispute. By this time, people had started to demand more empowerment of their decision-making capabilities. The creation of new institutional frameworks aimed to address the demands for spatial justice. In 1965 the National Office of Planning (ODEPLAN) and the Housing and

Urbanism Ministry (MINVU) were created. As part of the Housing and Urbanism Ministry, the Central Bank of Savings and Loans (SINAP), the Housing Corporation (CORVI), the Corporation of Urban Enhancement (CORMU), the Corporation of Housing Services (CORHABIT), and the Corporation of Urban Works (COU) were created.

This renewed institutional framework aimed to advance a more just urban life and to channel the radicalisation of the disputes over space. In this organisation of space, the role of urban practitioners was fundamental. Indeed, the disciplinary field of urban design was important for the political goals of Frei as well as for those of Allende. In the city the most visible expressions of political activity were concentrated, and it was considered one of the leading resources for social justice. During the times of Frei and Allende, the construction of housing in big cities was one of the most visible public policies, as well as the provision of new facilities for social activities. The priorities were transport, sanitation, public spaces, cultural facilities, and housing for the dispossessed. An interesting strategy involved building social housing in central areas of the cities, especially Santiago, aiming to reduce residential segregation and increase access to amenities and central areas for low-income communities (Figure 3.3). Reflecting backwards, 100 years to be precise, the ideas of the political class about urban development were widely different to the proposals of Benjamin Vicuña Mackenna in 1872. It was not segregation that was needed but integration; this exposed how this scientific urbanism had produced an efficient transformation in the mindset of Chilean authorities. The urban poor were no longer a problem to isolate and put away from the city centre but a group of dispossessed people who needed opportunities in the city to thrive.

Some of the emblematic housing projects built were Unidad Vecinal Portales, Unidad Vecinal República (1967), Unidad Vecinal Providencia (1968), and San Borja's urban renewal (Figure 3.1). Also, self-construction became an alternative promoted by the government. For example, in 1966 Frei organised a participatory process for designing houses for low-income and middle-class families in Santiago. The name of this initiative was Operación Tiza (Operation Chalk). It consisted of families drawing their 1:1 scale blueprints of dwelling units on the ground and then the state constructing the houses based on the people's design. This process was assisted by architects and builders, who guided people in the optimisation of the designs, for example distributing the rooms better and deciding on the materials for construction. However, beyond these good intentions, these operations contributed to consolidating new neighbourhoods in urban peripheries, instead of integrating them into the urban grid. Although innovative and participatory, the Operación Tiza represented an interesting exploration of solutions for building housing under a participatory scheme, rather than a robust public policy of urban development.

Figure 3.1 Remodelación San Borja, in Santiago's downtown. The picture is from
2015.

Source: Own elaboration.

Despite the institutional transformations, the illegal occupation of lands
multiplied (Figure 3.2); in 1969 in Santiago alone 35 new *tomas* appeared
(Gross, 1991, p. 43), and the innovation from the state was not enough
for the accelerated process of urbanisation that Chile was experiencing.
Empowered people and a frightened oligarchy were not a good mix. Political radicalisation, international interest in the raw materials available in the
national territory, and the rising inflation that had been plaguing the country
since the 1930s fostered social tension. Space was under dispute, and politics engaged to organise the demand of dispossessed people to ensure the
defence of private property on the one side and guide the struggle for spatial
justice on the other side.

In this convoluted historical moment, the political project of Salvador
Allende and the Unidad Popular (Popular Unity) proposed a pathway
for redistributing power and ensuring social justice. It was a revolution
with *chicha*[6] and *empanadas*, the so-called Chilean way to socialism.
Detached from the methodologies used by the Cuban revolution, Allende
bet that he could build a socialist state within the institutional frameworks

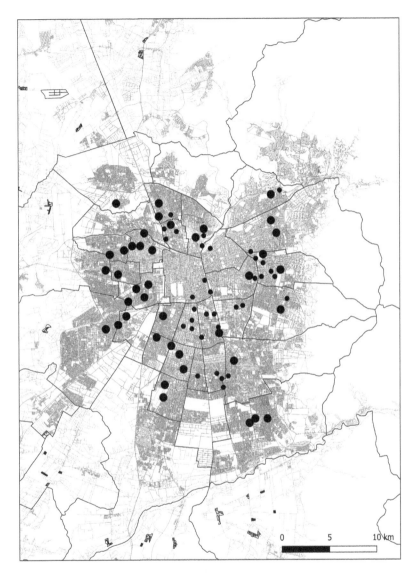

Figure 3.2 Localisation of *tomas* in Santiago in 1972.

Source: Own elaboration based on Castells 1987.

of the Chilean democratic republic, without using violence. In this goal, space was a key issue. The production of space and territorial productivity would be at the centre of the political transformations developed by Allende for building a fairer society. On the opposite side, the oligarchy started to organise its influences and political power to defend its position and property rights. For them, Allende represented a threat. In this dispute over defending/contesting property rights, the position of the military forces was fundamental. The contestation for land was not between equals. Allende received support from some prominent members of the military forces,[7] which lent security to the implementation of his programme. Nevertheless, within the army, the division was not even; oligarchical influence was always dominant.

Historically, evidence suggests that the military forces have been loyal to the oligarchy for defending property rights as the most sacred value of the nation (Salazar, 2003a, 2003b, 2009). Gabriel Salazar counts 23 times in which Chilean military forces have fired their guns against non-oligarchical Chileans, smashing every attempt to produce a fairer society (Salazar, 2012a). This is evidence that force – not justice – has been the engine of Chilean history. Allende was elected president by the majority and confirmed by the parliament, but not by the oligarchy.

In three years, Allende's administration achieved a comprehensive implementation of urban policies through the coordination of ODEPLAN (Oficina de Planificación Nacional or Planning Department of the Government) and the Urban Development Department. The state took part in urban development as one more member of the market. The government acted as a stakeholder, fostering a different approach to the city, starting with the empowerment of the grassroots in a well-intentioned bottom-up method of urban design. In Allende's programme, the objectives for urban development were increasing access to good housing, reducing spatial segregation, and using the land as an asset to redistribute wealth. Although approached slowly, community engagement was fundamental for fostering the revolutionary spirit of Allende's programme.

The plan was to build houses around job areas, namely Santiago's downtown (Figure 3.3). Several housing projects were designed and developed in Santiago, for example, in the neighbourhoods named Che Guevara (1970), Villa San Luis (1970), Cuatro Alamos (1971), Mapocho-Bulnes (1971), Plaza Chacabuco (1971), and Pozos Areneros (1971), to mention just a few (Table 3.3). These project typologies followed the international style, with four-storey buildings inserted in public spaces and with common facilities. These strategies, applied in Santiago and other cities, resulted in a significant decrease in the housing deficit, from 592,324 in 1970 to 419,000 in 1974 (Figure 3.4). However, these attempts to reduce segregation were not

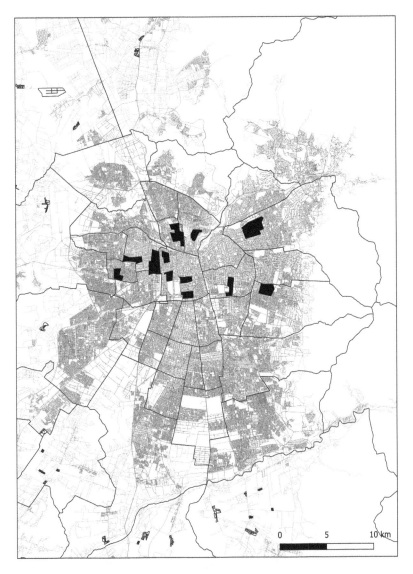

Figure 3.3 Areas of Santiago where social housing programmes were built between 1970 and 1973.

Source: Own elaboration.

Table 3.3 Social housing projects of Allende in Santiago.

Programme	Dwellings
San Luis	1100
Cuatro Alamos	778
Nuevo Horizonte	188
Salvador	2150
Mapocho-Bulnes	1200
Che Guevara	1490
Tupac Amaru	2267
Santa Monica	2500
Barrio Civico	150
Pozos Areneros	206
Ramón Allende	640
Plaza Chacabuco	510
Las Carabelas	250
TOTAL	13429

Source: Own elaboration.

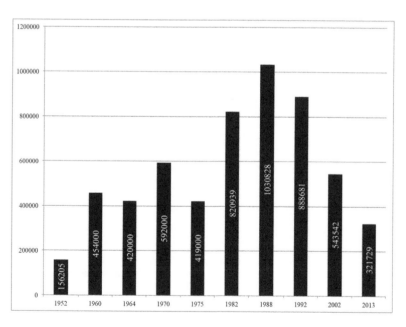

Figure 3.4 Housing deficit in Chile between 1952 and 2013. Allende's period reduced the housing deficit drastically, while the dictatorship of 1990 faced a deficit of more than a million housing units.

Source: Own elaboration based on Housing Ministry data.

welcomed by the oligarchy. Loyal to their tradition, they preferred to have the *rotos* living far away from the high-class houses. One of the emblematic cases was the Villa San Luis (Figure 3.5) in the heart of Las Condes district, a high-income area of the city. This new settlement provoked an awkward

Figure 3.5 Villa San Luis Building in 2015. The estates were sold to real estate companies that made a new financial district (Nueva Las Condes). The last resident of this villa left in 2015. This was the last bastion of the Allende administration's idea for integrating the poor and the rich via housing projects.

Source: Own elaboration.

reaction from the elite members who were concerned about losing the exclusivity of their neighbourhood (Gross, 1991, p. 44). The urban poor, following centres of opportunity, were already building informal settlements in Las Condes. Villa San Luis was a strategy for changing the status of this settlement from informal to legal housing estate. Also, it was an example of installing housing for low-income communities in high-income districts. In the view of the elite, the exclusivity of land in some areas of Santiago was contested not only by the urban poor but also by the government. The sacrosanct right to private property and the preservation of a still oligarchic landscape was under attack, and this was reshaping the urban space.

Both Frei and Allende's political agendas were facing demographic change and a society more aware of its rights, demanding better living conditions. Henceforth, their initiatives and transformations aimed to reduce the historical inequality and asymmetries of power that had characterised Chilean society since the foundation of Santiago in 1541. Of course, this was not a desirable outcome for the oligarchy so used to avoiding the urban poor and circulating among members of its own social class only. After these events and the social transformations towards a more just country implemented by the governments of Frei and Allende, the oligarchy was frustrated. For the first time its capacity for influencing the government was reduced to a minimum. Oligarchs were scared of the day when these masses of *rotos* would batter down their doors and take all their possessions, including all of their lands. Rather than being concerned with the politicisation of society, they saw the danger of the genuine democracy that was being forged, and it threatened their social status.

The solution was fierce and merciless, and as had happened 22 times before, the Chilean army aimed its guns at other Chileans to defend the private property of the few at the cost of the many. On 11 September, 1973, the democracy ended with the coup d'état led by Augusto Pinochet. Private property was safe again. Another revolution would start – a neoliberal revolution. Private property would return to being an untouchable right. The oligarchy unleashed violence to reinstall its hegemonic power, ensuring that it would control the economy, the government, democracy, freedom and – of course – urban society as a whole.

> Right-wing criticism, whether liberal or neoliberal, attacks urbanism as an institution but extols the initiatives of developers. This leaves the path open for capitalist developers, who are now able to invest profitably in the real-estate sector; the era of urban illusion has given them an opportunity to adapt.
>
> (Lefebvre, 2003, p. 163)

Concluding remarks

In this chapter, the research on urban design under neoliberalism has presented evidence that suggests its close relationship with private property rights. Also, the role of the elite in shaping the city has been situated as one of the main components that justify the emergence of urban-design-under-neoliberalism. The revision of the urban history of Santiago treated at a global level shows that private property and profitability were the main engine of Santiago's urban transformations and, therefore, the main criteria for shaping its urban form. Santiago's urban space has resulted from a historical domination of the land by the oligarchy, whose main goals has been extracting value from space: extractive activities in the territory (mining and agriculture), land transactions, construction, and urban development. This finding informs the roots of urban design under neoliberalism. The disciplinary field of urban design has been instrumental for ensuring the efficient exploitation of space as a resource for creating wealth. Also, the chapter exposed the historical neglect of the dispossessed by urban designers, which forced them to live in a constant struggle to finding a place in the city. With the exception of the period between 1938 and 1973, the urban poor have scarcely been considered when thinking of the future of the city. This chapter provided evidence to sustain the existence of a historical dependency of urban designers on the decisions made by an elite, evidencing a limited capacity for bringing plans designed for the sake of the common good into reality. All urban initiatives for enhancing Santiago were subjected to the interest of the ruling class.

Spatial dialectics as a method for articulating the historical analysis has served to understand how inequality – read as an historical condition in the city of Santiago – has been spatialised, defining clear areas for rich and poor. This process is developed in the following chapter, when during the dictatorship diverse actions were taken in order to accelerate the segmentation of the city based on household incomes. Wherever there is a space to dispute, there will be a dialectical method capable of investigating it. The global level of analysis employed in this chapter allows a valorisation of the historical dimension in order to understand the way in which a city has been formed by the reproduction of certain patterns of design based on profiting from spatial transformations.

Given that the rules of capitalism (profitability of actions and the rule of supply and demand) have shaped Chilean urban life, the defence of private property rights has become one of the main objectives of the ruling class. In this way, the role of spatial disciplines such as urban planning, architecture, urban design, and geography has been vital in generating strategies to defend the right to property through legislation but also through the organisation of the city. Just as Lefebvre points out, these disciplines played a vital role in preserving the hegemony of capitalism (Lefebvre, 2003), and in doing so they operated in a contradiction. Their ethics are cracked. The interests of capitalism drastically limit their role as designers for the betterment for everyone and they are easily co-opted by oligarchical interests. On the other hand, what I have shown is that people and not specialists have developed strategies for contesting private property for the sake of the common good that have actually transformed the urban form of Santiago. This fact implies that historically, urban specialists lacked fluid connections with low-income communities for interpreting their demands. The best urban designers for the urban poor have been, precisely, themselves.

In Chile, and particularly in Santiago, the political and economic power is in the hands of the land owners. This condition has remained the same from the beginnings of the city, and it explains the relevance given to defending this right either by law or by using military force. Therefore, democratic institutions are limited by the interest of the major property owners. The distribution of land for producing just spaces seemed difficult under the historical institutional structures reviewed because the oligarchy always made the final decision. Every time that a distribution of the way of using the territory occurred, the elite reacted with repressive strategies to defending property rights. For instance, the only time in history that a distribution of land for social wellbeing occurred was during the governments of Frei and especially Allende, but the result was a coup d'état and 17 years of a repressive dictatorship that developed a political constitution that ensured property rights prevailed. This is the current political, economic, and social system, and Chapter 6 delves into the features of what may be categorised as the neoliberal period of Chilean history. Neoliberalism in Santiago represents the concretised utopia of the Chilean hegemonic class, based on the subjugation of people to the rules of the market, property rights, and the rule of law.

Notes

1 The urban history of Santiago has been studied under the lenses of environment (De Ramón & Gross, 1985), architectural style (Boza, Castedo, & Duval, 1983; Eliash & Moreno, 1989; Guarda, 1978), hygiene (Pérez Oyarzun, Rosas, & Valenzuela, 2005), and the urban poor (Espinoza, 1988b).
2 In the original language: "El ejercicio del derecho de propiedad está sometido a las limitaciones o reglas que exijan el mantenimiento y el progreso del órden social, y, en tal sentido, podrá la lei imponerle obligaciones o servidumbres de utilidad pública en favor de los intereses generales del Estado, de la salud de los ciudadanos y de la salubridad pública".
3 Karl Brunner influenced urbanism in Chile and Colombia as well. "'[S]ocial urbanism'. What is it, exactly? The term itself is not new. It was coined by Karl

Brunner (1887–1960), an Austrian urban planner working in Bogotá in the 1930s. Brunner rejected the Beaux-Arts and later modernist utopian impulse of designing cities from scratch, and called instead for a practice that recognised what was already there. In that sense, he was a good seven decades ahead of the now orthodox attitude to the informal city" (McGuirk, 2014).

4 The main architecture journals in Chile were *Revista CA* (1968-present), *AUCA* (1965–1985), and *Arquitectura y Construccion* (1945–1950).

5 *Tomas* is the Chilean name given to community organisations that were illegally occupying land for building housing.

6 *Chicha* is the Chilean national drink.

7 René Schnider was the chief commander of the Chilean Military Forces, and he confirmed the military's commitment to defend Allende's project several times by saying that he would protect the constitutional presidency of Chile with his life. Schnider would die from a gunshot wound in 1970. A fascist fanatic killed him at the entrance to his house.

4   Neoliberal transformation of urban space in Santiago

Introduction

This chapter presents how urban design under neoliberalism was implemented. Thus, the findings explain how certain urban strategies that emerge from the set of principles of neoliberalism have been actually shaping the urban space of Santiago. By doing so, the chapter articulates the recent history of Santiago with its neoliberal transformations in relation to urban practices and political-economic goals. To do this, I analysed recent urban strategies that illustrate how urban design was aligned with the goals of the neoliberal project in Santiago.

The chapter contextualises how the neoliberal project of Milton Friedman and the Chicago Boys transformed the way urban development was conducted in Chile. After presenting some of the main discussions about the process of neoliberalising the city of Santiago, the chapter illustrates the main features of urban design under neoliberalism in Santiago using three urban strategies that well represent how neoliberalism has changed the urban design practice in Santiago. These strategies are the social housing policies, the real estate business, the public-private partnership, and, to narrow the discussion, I reflect on what I consider an urban icon of this period: the Costanera Centre building, a shopping mall, hotel, and office building in a central area of the city. These elements were selected to show how neoliberalism operates in different urban design typologies as I cover changes in space and policies and attempt to illustrate the decision environment for city making in Santiago. Before the conclusion, the chapter presents some practices that are contesting the ways of urban design under neoliberalism.

Free-market political economy in Santiago

In the case of Santiago, the process of neoliberalising the urban processes started in 1975. As a matter of fact, Chilean neoliberalism emerged from the operationalisation of a theory of political economy developed by Milton Friedman for the dictatorship of Augusto Pinochet (Daher, 1991; Salazar, 2003a; Solimano, 2014). On 11 September 1973, a coup d'état overthrew the democratic government of Salvador Allende. A dictatorship led by Pinochet governed the country for the next 17 years. Recognising the need for economic competency, Pinochet sought counsel from notable scholars opposed to Allende's ideas (Solimano, 2014). In 1975 he assigned ministries to a group of economists from the University of Chicago known as the Chicago Boys,[1] who implemented a programme of political economics[2] that came to be known as neoliberalism. The proposal aimed to reduce the size of the state and to increase reliance on free-market economics as the principal instrument for the development of the country (Arancibia Clavel & Balart, 2007). Pinochet accepted the ideas, and thus the Chilean dictatorship became a pathway to the liberalisation of the economy. The model argued that the free market would lead society to individual freedom, an idea that was presented to Pinochet by Milton Friedman himself. He went to Chile and met with Pinochet in March 1975 to become familiar with the Chilean reality in order to propose a plan, and this plan was delivered to the dictator on 21 April of the same year by Friedman in a personal letter. Friedman stated that the implementation of these reforms was urgent and that gradualism was not an option (Friedman & Pinochet, 1975), advocating for triggering the neoliberal revolution. Friedman was explicit: this was a shock programme that aimed to resolve the economic issues in months. Pinochet took the shock idea seriously and repressed fiercely all opposition to the programme.

There was a sharp switch in the dictatorship's policies towards creating the first neoliberal state in the world (Harvey, 2005). Since then, the state has encouraged private initiatives for providing social security and has advanced the privatisation of as many state functions as possible (Figure 4.1). A new era started, as private property and profit-driven decisions reshaped society (Atria, Larrain, Benavente, Couso, & Joignant, 2013). In order to perpetuate this transformation politically, a new constitution was promulgated in 1980 based on the supreme value of private property, strong restraints on the role of the state in fostering economic production, and a severe restriction on labour rights (Solimano, 2012). Furthermore, the constitution created the notion of a protected democracy by defining the military forces of Chile as the guarantors of the institutional order. Thus, any attempt to undermine its rules would be violently repressed.[3] Three components of this constitution had a significant influence on the creation of urban design under neoliberalism: the subsidiary role of the state, the defence of private property rights, and regulations that fostered entrepreneurialism and private initiatives. These transformations made economic growth the main

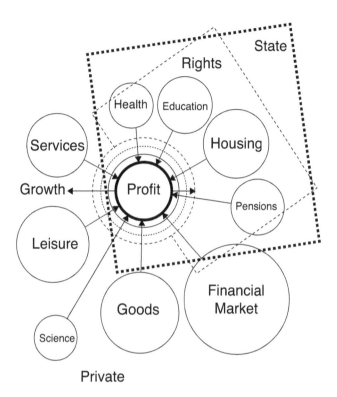

Figure 4.1 Diagram of privatisation as part of neoliberalisation. The whole society turns into a profit-oriented system, in which the state facilitates the transformation of rights into tradable services and goods. On the other hand, the main goal of the system is growth.

Source: Author.

goal of the state (Atria et al., 2013). Neoliberalism joined the political and economic realms of life, thus making sure that everything is understood on economic terms (Davis, 2017). In theory, the economic growth generated by consumption and productive activities would trickle down, reaching everyone and improving the lives of the people. As Devin Rafferty explains, the general idea of neoliberalism was to encourage companies and wealthy individuals to invest in creating jobs for people, whose salary would then be reinvested in the circuit of economic growth through the consumption of goods and services (Rafferty, 2017).

After decades of neoliberalism, per capita income increased significantly, but neoliberalism did not reduce inequality significantly (Figure 4.2); the distribution of incomes remained highly unequal (Gini Index), with no

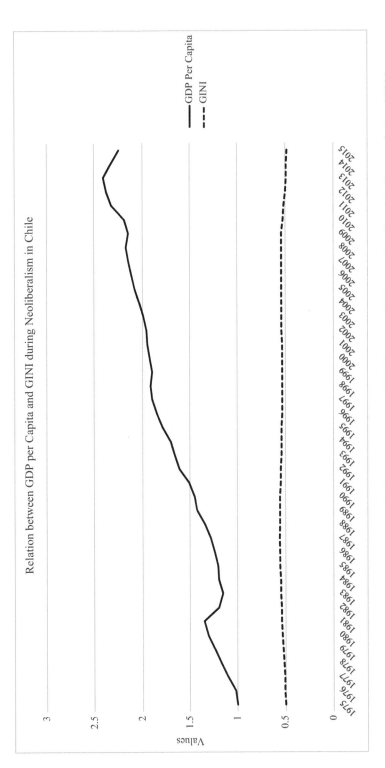

Figure 4.2 Relation between per capital GDP and distribution of incomes (GINI) between 1975 (year of neoliberal implementation) and 2015.

Source: Own elaboration based on Central Bank of Chile.

considerable changes having occurred since 1975. Neoliberalism has been efficient in generating growth but not in distributing it. Chilean inequality is a historical problem that neoliberalism did not alleviate at all.

The neoliberal transformation of Chile required a new urban development model, and thus the Housing Ministry and the Public Works Ministry were redefined and the urban planning apparatuses of the state were dismantled, with the government instead relying on private agents (real estate and urban developers) to define the urban form (Valencia, 2007). Santiago as a city under neoliberalisation process become a platform populated by urban components for speculation, with a civil society incapable of contesting a mercantilist notion of space (Rodriguez & Rodriguez, 2009). In the transformation of Santiago's urban development, the economist Arnold Harberger was fundamental. A professor from the University of Chicago, after marrying a Chilean he moved to live in Santiago, becoming a trusted advisor to Pinochet. In 1979, Harberger published a paper advocating the liberalisation of the processes of urban development as a strategy for advancing towards better access to housing. His argument was that great cities should grow by sprawl and that to promote this growth new regulatory frameworks were required (Harberger, 1979). Harberger's suggestions provided the theoretical basis for the creation of the National Policy of Urban Development on 30 November 1979 by Presidential Decree 420 (Daher, 1991), expanding the urban limits of Santiago (Figure 4.3). Also, this idea fostered land trading (Daher, 1991; Donoso & Sabatini, 1980; Gross, 1991; Trivelli, 1981). The policy presented four fundamental principles:

- Land is a non-scarce resource, therefore its use and value are defined by its profitability. It is subject to free trade, and restrictions on urban sprawl will be removed to allow the natural expansion of urban areas, following market trends.
- Housing scarcity will be relieved by private building companies, promoted by the state, but it is the responsibility of the market to deal with dwelling demand.
- Every improvement in the environment and in cities financed by the state should be oriented towards making land more profitable.
- In this policy, it is also stated that the goal of urban development plans is to improve the profitability of real estate.

(MINVU, 2014, p. 26)

Pablo Trivelli (in Rodriguez & Rodriguez, 2009) explains that in 1979 the National Planning Department implemented a series of policies to make urban development more profitable. These modifications caused the drastic expansion of the urban area by expanding its limits. Harberger's ideas suggested that increasing the number of plots available for construction should lower the price of land, creating more affordable housing close to downtown. For Antonio Daher (1991), Harberger's theory was completely incorrect because land can never be a non-scarce good, so it cannot be treated like any other commodity. Until 1984, this policy created massive flux in real estate development, mostly in land trade (Gross, 1991). Santiago increased its urban area by 62,000 hectares, but the price of land increased by 100% in two years (Trivelli, 1981). The theory of Harberger failed, and this pushed the dictatorship to develop a new policy. Thus, in 1985 a new National Policy for Urban Development attempted to amend the previous one, but it was never promulgated, serving only as guidance for good urban practices instead of serving as an obligatory regulatory framework (Gross, 1991; MINVU 2017). These transformations explain why after 1979 one of the main study subjects for urban disciplines was land markets.

In practice, these changes strengthened the position of the Chilean Chamber of Builders (CChC) in defining the urban form. CChC is an organisation that gathers real estate developers, building companies, and investors. Due to the nature of a market-oriented urban policy, those who control land prices and construction costs occupy an enviable position for defining how to develop the city. Specifically, CChC has become a gigantic union of real estate companies and builders. The richest families in the country, the most important urban developers, and the main financial institutions are members of this organisation, and their influence in politics is considerable. Table 4.1

Table 4.1 A selection of members of the CChC (Chilean Chamber of Builders). In parenthesis appears the ranking position of the wealthiest families in the country.

Banks members of the CChC	*Richest families members of the CChC (only the rank number)*
Banco BBVA Chile	Solari (1)
Banco Consorcio	Matte (2)
Banco de Chile	Angelini (3)
Banco de Crédito e Inversiones S.A.	Yarur (4)
Banco Santander Chile S.A.	Fernandez Leon (5)
Banco Security	Hutado Vicuña (6)
Bice Renta Urbana S.A.	Said (8)
CTI S.A.	Luksic (10)
Santander S.A. Administradora de Fondos de Inversión	Paulmann (12)
Mutuos Hipotecarios Cruz del Sur S.A.	Saieh (13)
Cencosud	Cueto (21)

Source: Own elaboration.

shows the financial institutions that are members of the CChC. Also, the richest families in the country are actively involved in CChC activities. The network of the power of the CChC is strongly linked with the interests of capital.

Because of its magnitude and resources, today in Chile the CChC produces large amounts of data to inform the processes of urban development. Since 1990, investment in construction has been boosted, increasing by 587% (Instituto Nacional de Estadísticas, 2017). In business, success comes from knowing customers' purchasing power and from targeting products. Therefore, creation of segments of the population based on their income facilitates the allocation of products (Silbiger, 2009). In real estate, this strategy was applied by classifying neighbourhoods by household income. Aware of this factor of efficiency in urban markets, the dictatorship produced a significant reordering of Santiago's neighbourhoods.

Using military forces, Pinochet reshuffled the urban segments, moving people from one side to the other as shown in Figure 4.3 in order to liberate areas for real estate investments. During the 1970s, the urban poor settled in strategic areas in Santiago, living in informal settlements, pejoratively named *poblaciones callampa* (mushroom towns), an inheritance of the *conventillos* of the beginnings of the twentieth century. In order to organise the city for market investment, Pinochet created 36 *comunas* (districts) segmented by social classes. This cleansing of the urban fabric for business may be summarised in three strategies:

1   There was a municipalisation of the city, a transfer of central powers to mayors in each of Santiago's *comunas*. It is worth saying that it was Pinochet who elected each mayor of the country; hence, the organisation of each municipality and its democratic empowerment was completely subordinated to the dictator's will. The mayors were there not to represent the people's voice but to organise the space for free-market urban development.

2   The National Policy of Urban Development 1979, although it did not declare its intentions, changed the idea of densifying the urban area of Santiago to fostering a strategy of urban sprawl. Harberger argued that all great cities experienced a period of urban sprawl (Harberger, 1979), so the policy and developers followed these ideas. In the new urban areas and districts of Santiago, the dictatorship allocated the urban poor space away from the centre.

3   Finally, there was an increase in the number of municipal divisions of Santiago (from 18 to 32), doubling the *comunas* and characterising each of them according to their socio-economic conformation. This socio-economic classification emerged from marketing theory, in which the marketer develops segments of the population in order to target its

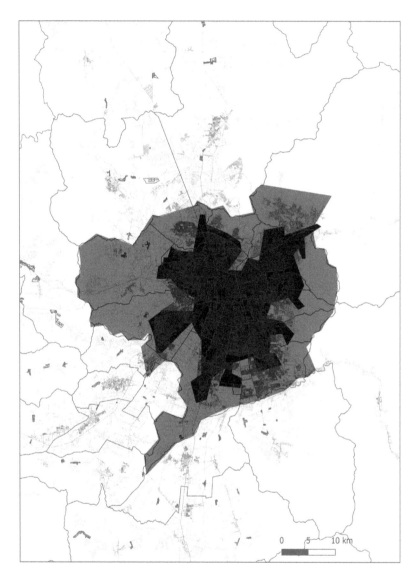

Figure 4.3 Santiago's urban expansion implemented by the National Policy of
Urban Development in 1979. The black line represents the urban limit
defined in the policy, the pale red blocks formed the urban area of
Santiago in 1979, and the black blocks are Santiago at the present.

Source: Own elaboration.

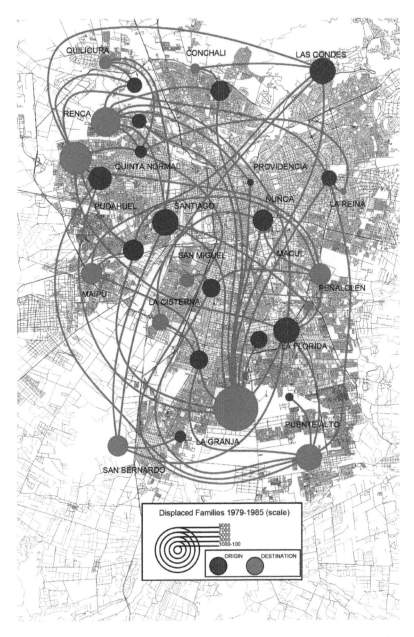

Figure 4.4 Displaced families during the dictatorship after the neoliberal implementation as part of liberalisation of land for real estate development.

Source: Own elaboration based on Aldunate, Morales & Rojas 1987. The source had another map; it was necessary to update the new districts of the city.

Table 4.2 Evolution of sale prices per square metre in Santiago from 1980 (neoliberal implementation) to 2016 (UF/m²).

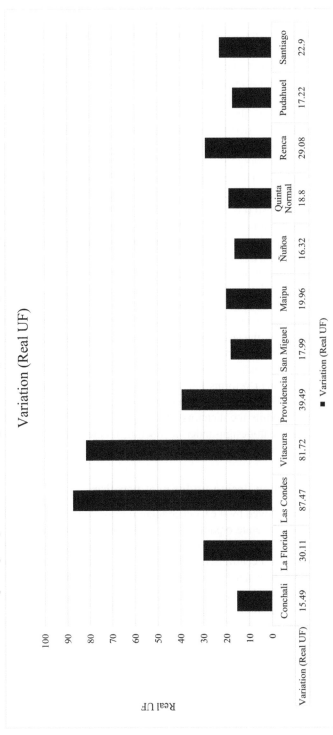

Source: Own production based on archives of *El Mercurio* newspaper in the section of "property selling", consulted in the National Archive of Chile.

products depending on consumers' interests, incomes, and desires. In this case, the segmentation considered spatial products (mostly housing) according to people's purchasing capacity. In other words, there was an organisation of the city by class.

As a consequence, the prices of housing increased drastically in some *comunas* while in others they increased at a low rate (Table 4.2). Therefore, Pinochet's urban plan for Santiago achieved the creation of communes for the rich (Las Condes, Vitacura, Providencia) and for the poor (Conchalì, Renca, Pudahuel). The segmentation of the city served to allocate adequate spatial typologies depending on their social class (revisit Figure 2.2). This transformation also produced a fragmentation between districts (*comunas*), lacking coordination on the broader scale of the city. Thus, each *comuna* defined its own rules and regulations, without the necessity of agreeing with those of other districts.

In 1988, opposition to Pinochet, organised as a political conglomerate and supported by international institutions and the United States, gained the chance of a plebiscite in order to decide whether Pinochet should continue as the head of the state. Pinochet lost the plebiscite, and the return to democracy started. After the elections of 1989, the Concertación de Partidos por la Democracia (Organisation of Parties for Democracy) ruled the country for almost 20 years. Once the country was a democracy, the expectations over urban development reforms had to wait, given the urgencies of a gigantic housing deficit and the shortage of funds available for infrastructural development. While the dictatorship set up the basis of neoliberalism, the outcomes of urban-design-under neoliberalism emerged during the transition to democracy.

Featuring urban design under neoliberalism

The model employed by the Concertación for resolving socio-spatial needs was based on the state organising social demands and the market providing solutions. This was developed by a system known as Mercado Publico (Public Market), a centralised Web-based platform through which the state demanded services and companies tendered to win the contracts. Thus, the state's choice was determined by simply hiring the company that provided the cheapest bid from among the companies that fulfilled the application requirements (Figure 4.5).

The state and the market speak the same language based on a set of financial methods under the cost-benefit approach. Both in private projects and in government-funded initiatives, the model of evaluation was based on the internal rate of return (IRR) and net present value (NPV). The IRR

Figure 4.5 Housing block in Bajos de Mena.
Source: Own elaboration.

measures the profitability of potential investments, considering the flows of cash from the beginning to the finalisation of a project, including financial costs and times. The IRR is complemented by the NPV, which determines whether an investment is risky and whether it is worth implementing in relation to the estimation of the inflation rate. Even if a project offers high benefits in relation to innovation, spatial quality, social value and cultural development – to mention just a few – if these features are not measured under a financial scheme they will not influence the evaluation of the project for public investment. As a result of this limitation, diverse urban projects are profitable but not beautiful, innovative, or socially embedded. The assessment of projects according to its social values is a useful instrument for rationalising public investment, but its weakness is that the instrument promotes reducing costs instead of increasing benefits, lacking comprehensiveness.

Chile's main goal during the 1990s was economic growth so that it could apply the trickle-down theory. This objective brought in new capital to invest in urban development, and many projects were reshaping the urban land-scape, but they were aiming for profit, not for producing a good city. A burst in construction brought more architects into the public sphere, recovering some relevance in the production of space. Mathias Klotz[4] enjoyed a leading position in criticising the urban face of Santiago. In 1993 he stated that 90% of the new buildings in Santiago would fail in any practical module in any Chilean architecture school (Klotz, 1993). Although his observation made sense, the critique lacked political understanding of the reasons for the selection of these projects, ignoring the influence of neoliberalism in producing those buildings. Klotz was separating architecture from politics and ideologies, when in reality it is embedded. Neoliberalism was invisible as a disciplinary problem for years.

In 2007, Felipe Assadi, former partner of Klotz, attempted to illustrate that good architecture is possible under the free market by designing a residential building downtown. He failed and then reflected: "Certain definitions of living standards have been deeply outlined by the real estate market. The definition of inhabitant preferences, income levels, regulations, and financial balances of developers describes a quite limited set of possibilities for the architect" (Assadi, Pulido, & Zapata, 2008, p. 48). Assadi recognised that "design on an architectural level almost does not exist, unless in ornaments – if there are resources for it – or in irrelevant elements, that do not involve real project decisions" (Assadi et al., 2008, p. 50). The IRR and the NPV were the mantras for making decisions on urban-design-under-neoliberalism. The next part of the chapter illustrates the practices of urban design under neoliberalism; we need to discuss the three realms of action based on social housing, real estate development, and public-private partnerships, which are the areas in which capitalists and urban designers have more visible effects on the city.

Social housing under neoliberalism

In 1990, Chile had a deficit of 1,030,828 housing units (Gilbert, 2000), a problem which demanded urgent solutions. The administration of Patricio Aylwin developed an ambitious social housing plan for building more than 250,000 new dwellings by 1994 (Rodríguez & Sugranyes, 2005). In order to do so, the state would hire private construction companies to build the dwellings required. The hiring of companies would be undertaken through a system of bids. Thus, the private sector, using public funds, would ensure profitability by building social housing. The plan used subsidies for housing, progressive housing programmes, basic housing construction, and a special programme for workers. Following the logic of the IRR and NPV, the projects offered an interesting rate for building companies, a low level of regulation, and large amounts of funds in order to provide a massive number of housing units in a short time period. However, this acceleration provided wealth for the few and new problems for the many. An emblematic example was the case known as Casas Copeva. These social housing estates were Volcán San José 1, 2, 3 and 4, built by COPEVA between 1995 and 1997 in Bajos de Mena in the southern area of Santiago. The project was funded under the scheme of the Basic Housing Programme and considered the initial construction of 2,306 housing units between 1994 and 1996, which implied significant relief to all those families that had previously been living in self-built shacks.

Casas Copeva flooded after the first heavy rain of 1997, and 600 families abandoned their recently inaugurated dwellings. The floods occurred because of the deficient design of the houses and a drastic reduction in construction costs. Loose regulations in relation to construction ignored the fact that private companies would aim to increase their earnings by cutting costs, equalling high profits for the company but low spatial quality. Furthermore, the oligarchical tradition of the Chilean elite was another problem in this model of development. Edmundo Hermosilla was the Housing Minister when, in 1994, the Housing Ministry purchased the land of Bajos de Mena from Francisco Perez-Yoma, who was the owner of COPEVA, a big Chilean building company. After purchasing the land, the state decided to construct social housing (Volcán San José 1, 2, 3 and 4) in Bajos de Mena. The building bid was assigned to COPEVA as well. A great conflict arose because Hermosilla and Perez-Yoma were good friends. Indeed, a newspaper discovered that Perez-Yoma gave Hermosilla a horse as a gift at the same time that the Housing Ministry was dealing with the purchase of the plots of land in Bajos de Mena. The connection between Hermosilla and Perez-Yoma was confirmed by the minister, and he resigned after the scandal, without providing an explanation of the relation between the horse and the purchasing of these plots of land from his friend. Moreover, COPEVA filed for bankruptcy and never paid for the deficient Casas Copeva. In the end, the Housing Ministry paid £2,000,000 to repair 600 houses. This is an example of how the business was profitable for the private company, expensive for the state, and a big problem for dwellers.

The accelerated production of housing did not consider the creation of friendly environments, urban connectivity, and amenities (Figure 4.6). Despite the accelerated production of housing, inequality on spatial outcomes increased. Between 1988 and 2002, the housing deficit was reduced by 50%, but the issue of the spatial quality of dwellings was not tackled, and neither were questions about the quality of urban life. Alfredo Rodríguez and Ana Sugranynes (2005) named this contradictory phenomenon as *los con techo*:[5] people owning their houses but living in precarious neighbourhoods, lacking connectivity to urban centres, facilities, and services, and suffering from a clear ghettoisation of their social spaces. At the beginning of the 1990s, urban design under neoliberalism followed an abstract approach to

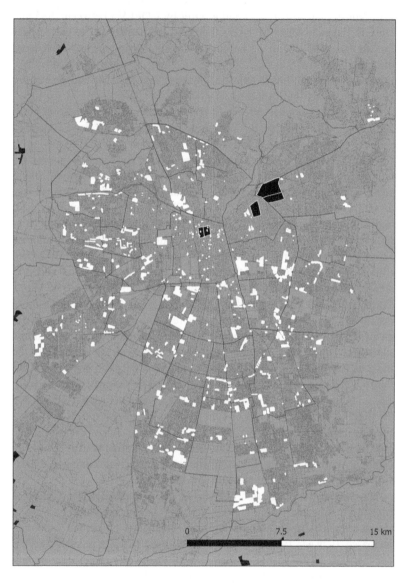

Figure 4.6 Map of Santiago marking the location of social housing projects (in violet) and the concentration of job places per *comuna*.

Source: Own elaboration based on CASEN 2015, MINVU, and Encuesta Nacional de Empleos 2017.

the urban that ignored the rich complexity of everyday life. More houses were built but not better urban spaces. Although it should be better to live in a cheap house than in a shack, the absence of reflective and comprehensive urban development turned the social housing projects of the 1990s into provisional solutions, and the state had to invest again in order to make changes.

Social housing policies produced under a scheme of the subsidiary state following financial criteria are conflictual in building better cities when regulations are loose or market-oriented. The model of subsidies in Chile allows the government to provide funds to families in order to complement their purchasing capacity and enable them to acquire a house through the market. The state finances the projects, private companies build housing, and citizens receive their own property while companies obtain their earnings. These companies learn to increase the profitability of their activities by reducing construction costs and building on cheaper plots of land. In Santiago, the cheaper plots of lands are very far from the centre of the city and away from job opportunities (Figure 4.6). As Camila Cociña explains, these conditions become more problematic under the logic of economies of scale, which foster the construction of large numbers of houses in vast areas of the city in order to maximise the utility of the investment (Cociña, 2012).

One of the causes of this problem was the goal of reducing the housing deficit over a short time period (eight years). This goal was an abstraction of the problem, focussing only on the mass production of dwellings and ignoring the comprehensiveness of urban development. Between 1990 and 2005, 2 million dwellings were built, reducing the deficit considerably (Rodríguez & Sugranyes, 2005); however, the positivist approach to analysing the social housing question then required a new scope. For Rodriguez and Sugranyes, the companies providing construction services to the state were very satisfied, facing no risks or pressures: The Ministry of Housing and Urbanism provides subsidies to people and building companies to construct the solutions. At the end of the year, the state gives back to these companies 65% of their taxes for construction costs. The state is protecting not only the companies but also the financial market by agreeing to finance the loans of people applying for these subsidies. MINVU pays the banks for people's credit insurance and assumes the responsibility in case debtors cannot pay their debts. There is no risk or competition. In this captive market, building companies do not need to consider new ideas, contributions, and practices. Neither the ministry nor the companies needs to open a debate over the social and urban costs of this massive production of social housing, including the costs of allocating services and amenities in the periphery (Rodríguez & Sugranyes, 2005).

In 2006 the approach changed with programmes like Quiero Mi Barrio.[6] This new approach represents a new urban design under neoliberalism, a phase of neoliberalism with a human face (Hidalgo Dattwyler, Christian Voltaire, & Santana Rivas, 2017). In this approach, urban-design-under-neoliberalism is employed to broaden access to private property, thus allowing low-income groups to access credit and debt. Thus, neoliberalism adopted a human face using urban and social housing policies, as a representation of a new approach to social development which incorporates values such as equity, social mixture, and sustainability. However, this is an ideological rhetoric strategy. At the end neoliberalism still expels low-income communities to new urban peripheries (Dattwyler et al., 2017).

Specifically, social housing in the neoliberal era of Chile increased the importance of urban development as an economic activity with the goal of reproducing capitalist modes of production. Social housing, by providing private property, contributes also to strengthening the importance of financial agents and the role of banks in the production of space.

Neoliberal real estate and the financial realm

It is a fact that during its neoliberal period, Chile increased the GDP per capita of the country to over USD 23,000,[7] reaching levels of purchasing power similar to those of Greece or Croatia in 2016 (World Bank, 2016). In addition, the access to financial credit and the promotion of private property rights increased the possibility of owning a house.

In 1973, 57% of dwellers were homeowners in Santiago, while in 2010 the percentage of homeowners in Santiago reached 80% (Simian, 2010). These conditions increased the importance of financial institutions in the housing market. Ivo Gasic Klett (2016) revealed that between 2010 and 2015, 40% of the acquisitions of urban land were by financial institutions, while only 31% were by real estate companies. For Gasic, these data expose the existence of an advanced process of financialisation in the housing market. The financialisation of housing implies the increasing dominance of financial actors and their practices, measurements, and strategies over the structural transformations of the mode of producing housing (Fernandez & Aalbers, 2016). For Raquel Rolnik (2013), this process has two main implications: first, the promotion of homeownership strengthens individualism and a consumption-based society; second, the financialisation of housing requires a more accessible mortgage system, lowering the requirements for credit in order to ensure that everyone is able to own a property. Rodrigo Cattaneo Pineda (2011) explains that after the reform of the Chilean capital market (MKI) in 2001, new mechanisms facilitated the capitalisation of companies, intensifying the relationship between financial institutions and real estate development. The reform fostered the diversification of companies' investments. For example, real estate developers received fresh funds from different financial resources, adding dynamics to the construction industry. Several financial firms acquired traditional real estate companies such as Moller-Perez Cotapos (Citigroup) and Penta Group (Security), and real estate developers such as Salfacorp S.A., Pazcorp, and Socovesa went public on the Chilean stock market. These new investors organised packages of investments to develop specific urban products such as pharmacies, shopping malls, shopping strips,[8] gated communities, and high-rise residential buildings. For Cattaneo, "the metropolis is not only a centre for articulating global capitalism. Materially, the fabrication of its spaces is one of the most essential strategies of adding value to capital" (Cattaneo Pineda, 2011, p. 8). The aesthetics of this process of financialisation were presented in a book titled *Infilling*, an apologetic exposition of Santiago's transformation by urban design under neoliberalism. The book describes how the consumers of dwellings prefer to live within the limits of Américo Vespucio (a circular road that surrounds Santiago) in consolidated neighbourhoods, making the city grow skywards. For Ivan Poduje, Nicolas Jobet, and Juan Martinez (2015), infilling is explained by the demographic transition of people aiming to live nearer downtown and their workplace in order to reduce commuting times, the fear of robbery (more common in houses), the increased prices of land, and the transformation of public policies that promoted the development of denser cities.

For instance, Jorge Vergara (2017) recounts that 40.6% of the residential buildings constructed between 1990 and 2014 had five storeys or more. Despite some level of appreciation of the infilling phenomena, Poduje et al. (2015) recognise the imperative of improving the coordination between authorities, companies, and civil society. Furthermore, most high-rise residential buildings are in the richest *comunas* of the city: Santiago, Las Condes, Ñuñoa, Providencia, and Vitacura (Vergara, 2017). For Jorge Vergara, the increasing number of high-rise residential buildings illustrates the verticalisation of the city. The financialisation of housing leads towards a process of infilling and verticalisation (Figure 4.7). Infilling is the process of populating central areas of the city with new real estate development and increasing the density of the district (Gatica, 2011). The financialisation is more complex.

Under financial logic, the way of defining where and how to construct housing depends mainly on the rules of supply and demand and profitability (Silbiger, 2009). Hence, investment in the richest *comunas* is prioritised because these areas are less risky for capital. The richest *comunas* of Santiago have consolidated housing markets, and the profit is secured. On the other hand, the poorer *comunas* receive projects with a lower quality of design because the customers have less purchasing capacity. Under the logic of supply and demand, if a *comuna* has reduced purchasing capacity,

Figure 4.7 Picture of Santiago's downtown and its verticalisation.
Source: Own elaboration.

the market provides cheaper housing. Between 1980 and 2016, the price of construction in Santiago varied by 15%, while the selling prices of housing increased by 242% (Figure 4.8). Additionally, the financial evaluation of real estate projects between 1980 and 2016 indicates that their profitability increased by 448% (Table 4.3). Real estate and construction are highly profitable businesses today.[9] Under this scheme of real estate as a financial asset, the only rule is obtaining a high rate of IRR in order to maximise profits in the short term. For instance, it is desirable to have an IRR of between 12% and 20% for real estate projects. The IRR over three years for five different projects in Santiago (middle-class *comuna*) and San Joaquin (low-income *comuna*) reveals that the IRR was 86% and 51% respectively, representing an outstanding revenue for investors.

In Santiago, the role of banks in real estate development is to provide loans for both builders and buyers. In the long term (20 to 30 years), banks receive a double revenue from the same spatial product, plus interest and the certainty that their portfolio of clients will increase. Construction and real estate are fundamental for banks. For instance, in 2012, 38% of the

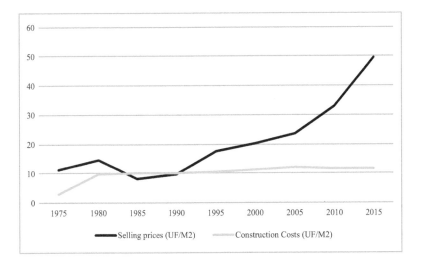

Figure 4.8 Construction vs. selling costs in Santiago between 1980 and 2016. Prices expressed in UF/m². Construction costs increases according to the CPI and are organised by the Housing Ministry, while the selling prices were obtained from the housing market published in El Mercurio.

Source: Own elaboration based on Ministerio de Vivienda (Tabla Unitaria de Costos Clase E), El Mercurio (Historical Archives).

loans from Chilean banks were used for construction and housing activities, including mortgages. The financial system has adopted a goal of increasing access to credit. One of the strategies of the Chilean Central Bank was to reduce the interest rates for mortgages in order to ensure that more people had access to credit. However, after the 2008 crisis, the banks changed the policy and started to fund a maximum of 90% of the total cost of a property. Nevertheless, interest rates in mortgages continued declining (Figure 4.9).

Another illustrative indicator of the importance of financial institutions in the real estate market emerges when we compare the number of houses sold to the mortgages granted. In this case, between 2005 and 2016, the number of mortgages granted was higher than the number of housing units purchased (Table 4.4). In other words, credit and debt have become the main instruments for acquiring housing.

Costanera Centre: the icon of the neoliberal city

After 2014, the postcards of Santiago started to present Costanera Centre as the one iconic building considered one of the main attractions of the city,

Table 4.3 Comparing the profitability of high-rise buildings between 1980 and 2016. "M^2 on Sale" is an average of the area available for selling with high-rise buildings of 18 floors whose typology existed in 1980. As a result of this hypothetical analysis, the variation on the profitability of the same typology between 1980 and 2016 was 448%.

Item	Data for 1980	Data for 2016
Construction Costs (UF/m^2)	9.9	11.71
M^2 on sale	24,063	24,063
Sales value (UF/m^2)	14.51	49.66
Total Sales (UF)	349,154.13	1,194,968.58
Initial Costs (UF)	*−238,223.70*	*−281,777.73*
Earnings 1st year	69,830.83	238,993.72
Earnings 2nd year	139,661.65	477,987.43
Earnings 3rd year	139,661.65	477,987.43
IRR	**19%**	**107%**
Variation of Profitability	**448%**	

Source: own elaboration.

Table 4.4 A balance between housing units sold per year and mortgages granted.

Year	Housing sold	Mortgages granted	Difference
2005	55	77,725	*−22,725*
2006	58	30,856	27,144
2007	59	32,688	26,312
2008	52	22,227	29,773
2009	57,263	13,098	*−79,835*
2010	46,872	145,276	*−98,404*
2011	56,865	70,738	*−13,873*
2012	67,135	86,866	*−19,731*
2013	69,007	52,181	16,826
2014	63,981	56,492	7,491
2015	81,194	36,263	44,931
2016	51,549	16,57	34,979
Total	**717,866**	**764,978**	*−47,112*

Source: Own elaboration based on the Central Bank of Chile, SBIF, and CChC.

although most urban designers stated that this skyscraper was problematic (Figure 4.10). Costanera Centre is a multi-functional building containing a shopping mall (Figure 4.9), parking, restaurants, and offices. It is the highest building in Latin America, at 300 metres. This building is the result of a personal project of a billionaire, Horst Paulmann, who implanted this tower as part of his legacy to Santiago.

Interest Rate for Mortgages

Figure 4.9 Interest rate for mortgages. If the rate is constantly decreasing, it means that the banks are reducing the requirements for granting mortgages, which is also a symptom of a real estate bubble.

Source: Own elaboration based on SBIF.

Figure 4.10 Postcard of Santiago. On the left is the Titanium Tower and on the right is the Costanera Centre Tower. At the back, you can see the Andes.

Source: http://isolatek.com/wp-content/uploads/2016/05/Torre-Titanium-Gran-Torre-Santiago-CHILE-4.jpg.

Regarding the findings of Chapter 5, Costanera Centre is an icon of how land tenancy and wealth are still the vital sources of power in Santiago. The history of the Costanera Centre started around 1988 with the first sketches made by the architecture studio Alemparte y Barreda. The studio proposed the idea to Horst Paulmann, seeking funds, and he proposed inviting Cesar Pelli to be part of the design team. Alemparte y Barreda agreed. Thus, in 1991 Paulmann bought a plot of land and subsequently purchased other plots nearby in order to accumulate space to build his own Tour Eiffel.[10]

However, his building had more complexities than just its aesthetic considerations. For instance, the Costanera Centre transgressed transit rules, and Paulmann used his influence to change the regulations that jeopardised his financial goals for this project. For example, the municipality of Providencia ignored the construction of this tower even when its permits had expired. The building's construction authorisation (which in Chile is granted by local government through its Municipal Department of Construction) expired in 2006, but the work continued after this date. Indeed, the National Comptrollership of Chile determined that the Costanera Centre was under construction with an expired permission (Dictamen 27.392 of 18–06–2007). However, instead of stopping the work, the building company was given the chance to regularise its construction, and the municipality had to amend the permission in order to legalise the situation. Moreover, new irregularities in this project continued to emerge (Fossa, 2011). Patricio Herman (2013) says that the only reason that projects of such impact as Costanera Centre are built is because of the lenient attitude of the authorities towards building companies.

In Chile, all urban projects that are considered to be high-impact construction have to apply for a Certification of Environmental Impact and an Urban Transport Impact permission. Applications for both permissions are evaluated and eventually granted by the central government. However, evidence collected by Patricio Herman showed that Costanera Centre was constructed illegally. For instance, the environmental permission was obsolete because it was granted on a previous version of the design whose area was 30% smaller. Similarly, the Urban Transport Impact Study approved this project, despite the fact that the building lacked a strategy for reducing traffic, and the construction continued.

A third irregularity is that the area on which the building was constructed was composed of several plots of land whose legal fusion was not certified when the works were started (Cociña, 2007; Herman, 2013). Eventually, by the end of the construction, the project had obtained all of the necessary permissions.

Costanera Centre illustrates how the good city is threatened by urban-design-under-neoliberalism, because the regulatory frameworks are shaped for the sake of capital and not for the sake of better space. Also, this project shows that when regulations jeopardise private investment, the authorities may twist the rules, change them, or simply turn a blind eye. From the perspective of an urban designer, the project rejected the potentialities of pedestrian areas and, instead, was placed in one of the most complicated areas in the city from the perspective of traffic and the urban grid. An intricate space with irregular forms, it is difficult to access because even before the construction of this building the area suffered from high congestion. The Costanera Centre shopping mall creates an illusion of a public space on the inside. The sidewalks that surround the building are inhospitable and vast and lack design. Inside the building, the pedestrian paths are luxurious and smooth, with an impressive void in the centre and escalators composing a scene of desirable consumerism (Figure 4.11). The private space inside the Costanera Centre creates a mirage of public pavements in order to invite pedestrians into the consumption temple where they can worship the shops. Protected from the precarious urban design proposal of the outside, the interior seduces through a vibrant space.

One of the most interesting aspects of urban design under neoliberalism is its capacity to adapt to times and changes in consumption patterns. As Liliana de Simone (2012) revealed, other shopping malls such as the Parque Arauco were initially designed as an introverted box of bricks but are now more an open space, aiming to create the illusion of a public space. The same happened with the initial form of Plaza Vespucio, another emblematic shopping mall in Santiago that today operates as a civic centre by accommodating banks, restaurants, and even areas with public services. Urban design under neoliberalism is adaptable to changes in *consumption patterns*, and it recognises the preferences of people when they change their behaviour. As the biggest building ever constructed in Chile, the Costanera Centre is an icon of the neoliberal era of this city. It is not only a temple of consumerism but also a representation of the persistent hegemonic power of land owners in the process of shaping Santiago. In this case, the land owner decided to build a gigantic tower, and the authorities, instead of stopping him, preferred to facilitate the process of construction. Of course, from the perspective of a politician, this huge project created 3,000 jobs during its construction and promised to create 8,000 jobs after its inauguration. Horst Paulmann and his tower may be the symbol of urban design under neoliberalism. What is interesting is that the tower is facing a crisis because it has not been capable of fulfilling its profit expectations in the short term and has had to readjust its plans for the near future.

Contestations against urban design under neoliberalism

Urban design under neoliberalism has produced several problems that have triggered reactions, mainly from academia and some practitioners, fuelled

Figure 4.11 Inside Costanera Centre and the scene of the stairs and consumption in process.

Source: Own elaboration.

Figure 4.12 The tower of Costanera Centre has been popularly named MORDOR
in reference to *The Lord of the Rings* movies, where a gigantic tower
with an eye at the top represents the evil of the Middle Earth.

Source: Own elaboration.

by local communities aiming to contest the way that the city was producing unfair outcomes. Transformative urban practices emerged to provide
different solutions for the city, using various means: constructing collective housing, questioning the city by occupying public spaces, visualising
problems as a consequence of the neoliberalisation of spaces, and seeking
institutional responses.

These processes of contestation are optimistic because they not only
criticise the urban form and its modes of production but also generate practices of change. Indeed, there is a latent revolutionary potential expressed in
these approaches that may lead to a more extended urban revolution.

Beginning in 1987, a group of neighbours of Estacion Central organised
themselves under the name of UKAMAU to claim their right to housing.
After two decades of struggling against the rigid regulatory framework of
the state and protesting over their rights, in 2011 two architecture students
from ARCIS University proposed to their tutor that they use the demands

of UKAMAU as a case for a practical module. The tutor was Fernando Castillo-Velasco, one of the most influential architects in the field of social housing, characterised by fostering community engagement in the production of spaces.

Since the 1960s, long before Alejandro Aravena's Elemental, Fernando Castillo-Velasco had designed social housing of high quality in Santiago and other cities. Thus, Castillo-Velasco said to his students that not only would they design housing for UKAMAU but they would also build it. The design considered 424 apartments of 62 m² and significant areas of public space near Santiago's downtown. In this case, the process was bottom-up: the people of UKAMAU designed the project and found the plot of land for its construction. They came with the finished project to the Ministry of Housing to negotiate its construction, under UKAMAU's terms and not considering the possibilities of the market or solutions provided by the state. The young architects of ARCIS were hired by the state to develop the project, while UKAMAU became a construction cooperative that would be hired by the state to construct their houses.

The model of UKAMAU has been replicated by other communities in the country (Pudahuel, Cerro Navia, Pedro Aguirre Cerda, Antofagasta, Calama), becoming a practice capable of, at the least, empowering people to subvert urban design under neoliberalism. Similar to UKAMAU are the Movimiento de Pobladores en Lucha (Movement of Struggling Dwellers) and ANDHA Chile, both grass-roots movements organising social forces in order to claim their right to housing. Furthermore, different organisations have extended the idea of housing as a right to foster the reclamation of the city for the people.

The public space has remained as an open space for free expression since Chile returned to democracy in 1990, but even today repression is applied by the state to marches and special types of demonstrations. Despite the supposed freedom of speech, it is constrained if it becomes awkward for the authorities. Thus, the arts are still an apparent inoffensive occupation of the public space, but they still may become a provocative action that awakens the consciousness of citizens and makes them question the way that they live under neoliberalism. This is the case of Proyecto Pregunta, an ephemeral intervention designed by the collective MilM2, whose aim is to enable debates through the activation of public spaces.

> Proyecto Pregunta (Question Project) is a tool for community engagement and participation designed to foster the collective generation, visualisation and viralisation of debates on the public space. Proyecto Pregunta is a critical device, aimed to intervene with civic participation

methodologies in order to question and open new types of community awareness and civic engagement.

(Boano & Vergara-Perucich, 2017, p. 145)

Asking the citizens what they feel and think about their urban lives has been a successful initiative for activating the public space. Indeed, the project has now been performed in thirty different spaces in Chile, producing more than 2,000 questions. Also, Proyecto Pregunta has been implemented in Europe and Brazil. More ironic are the activities of Grupo TOMA, a collective of architects and urbanists whose goal is to generate collective reflections on the urban space, occupying it and encouraging people to question the city by using it. They believe that cities are political and that each space has a political voice that needs to be heard.

> By making these territories "speak", we intend to considerate the political role of architecture as quotidian activism. It is an attempt to understand the different means through which architecture is able to politicise spaces and reflections and to identify disputes and conflicts. In this retrospective reflection, biographical, we have used our own experiences, but the discussion remains open to be redefined as new aspects and elements of urban transformations in the Neoliberal context will continue to appear.
>
> (Grupo TOMA in Boano & Vergara-Perucich, 2017, p. 170)

Grupo TOMA considers that objects and ephemeral interventions in the public space are a way to activate a series of reflections on the city's history and also on the ethical role of architects. Their practice valorises the utopian thinking of a more collective city in which the role of the architect is precisely to facilitate the encounters of people and to engage in the pursuit of a new identity. It is possible to say that for Grupo TOMA the neoliberalisation of cities in Chile has erased the identity of its spaces, and they are trying to reclaim it by using even neoliberalism itself to create identity. For instance, in the Chicago Architecture Biennial of 2016 the group performed a project named Especulopolis, an open invitation to speculate with the urban space. This project was searching for the traces of urban neoliberalism following the lead of speculation, proposing a transparent exposition of how cities have been manufactured following the ideas of the Chicago Boys in Santiago.

Along with the contestation emerging from grassroots movements and activists, the state has also provided responses to the problems generated by urban design under neoliberalism. After several years reproducing this model, the state now is aiming to provide a different approach to the problem of housing for the poor. In 2013 a group of specialists started the design of a new National Policy of Urban Development, aiming to tackle these problems and finally replace the policy of 1979. The formulation of the new policy was published on 4 March 2014, with significant transformations to the way that urban development was to be conducted, taking a step away from neoliberalisation. The goals are ensuring the production of more equitable cities and more socially integrated and more democratic urban environments (CNDU, 2015). It is focussed on people and their living conditions, promoting a sustainable approach to urban development, valorising the public sphere, and assuming a progressive implementation of its guidelines and strategies (CNDU, 2015). This policy presented five pillars for action:

1   Social cohesion policy: guarantee equal access to public goods, stop and reverse social segregation in cities, reduce the housing deficit, develop a land policy for fostering social cohesion, promote community engagement, increase connectivity and universal access to cities, and incorporate remote urban settlements for policies of social cohesion.
2   Urban economic development policy: create urban conditions for fostering economic development and innovation, integrate urban planning with investment programmes, amend and organise the land markets, improve the competitiveness of cities in the light of globalisation, improve urban planning instruments, monitor the efficiency of infrastructural development, rationalise transportation costs, and flexible planning instruments for changing contexts.
3   Environmental urban development policy: incorporate eco-systems into urban planning, incorporate disaster risk reduction, manage waste and natural resources efficiently, monitor urban environmental factors, foster sustainable uses of urban land, foster the use of bicycles and promote the pedestrian use of spaces.
4   Identity and heritage policy: valorise the built environment as part of the identity of communities, recognise and protect the cultural value of the built environment.
5   Urban governance policy: decentralise urban management, organise territories into four scales of governance (national, regional, metropolitan, and municipal), implement comprehensive urban planning and binding participatory processes, index urban quality developments to ease the measurements, and accelerate the approval of plans and projects for urban development.

The results of the policy were not completely satisfying for all specialists. For example, Dr Jorge Inzulza (in Lopez, Jiron, Arriagada, & Eliash, 2014) exposed the lack of coordination between the proposals of this policy and the planning instruments in each regional context. Chile has quite a diverse geography, from the driest desert in the north to evergreen forest in the south, which represent very different modes of the production of spaces. Javier Ruiz-Tagle (in Lopez et al., 2014) criticised the lack of clarity in some of the guidelines, leaving space for interpretations that may differ for the development of just spatial transformations. For Ernesto López-Morales (2014), the exclusion of people from the amenities of urban centres was not completely tackled by the proposal.

This policy provides a new approach for fortifying the role of the state, but it is not yet clear how the policy contributes to dismantling urban-design-under-neoliberalism. As of this writing, the policy has not yet been implemented, so it is not possible to evaluate its results. It seems like an adequate response, although the Chilean Chamber of Builders has already rejected part of its proposals. Considering the power this organisation has in matters of urban design, although the new policy seems promising it would be prudent to preserve some scepticism about its outcomes at the beginning.

Concluding remarks

The chapter presented how a series of neoliberal political-economic trans-formations reshaped the way of practising urban design in Santiago. It has illustrated how urban design under neoliberalism operates in the produc-tion of the urban space, characterised by chasing profit-oriented goals and also by embracing economic growth as the main objective for spatial trans-formations. The exploration of how urban design under neoliberalism was implemented reveals that the transformation of the urban space of Santiago under neoliberalism occurred under a profit-oriented scheme, employing economistic theories for making decisions. Hence, it is possible to argue that urban design under neoliberalism adopts borrowed theories from eco-nomics, embraces a positivist reasoning, and neglects its capacity for imag-ining cities beyond the limits of capitalists' interests.

These criteria were institutionalised by the state in relation to most of the processes for assigning funds for urban projects. This means that not only private actors but also the state aimed for profit-oriented goals. This hap-pened because both public and private realms assumed that growth and eco-nomic development were the best way to modernise the country. Analysis of social housing developments from 1990 until 2006, along with the public transport system, urban highways, and permissions for construction, shows that the urban apparatuses of the Chilean state have a clear leaning towards making decisions for the sake of capital instead of actually protecting the public interest. However, this particular observation is nuanced because the Chilean Constitution and the objectives of the state follow a neoliberal interpretation of reality: if capital is healthy and growing, then there are more funds for investing in social programmes because such growth increases tax revenues. This is the trickle-down theory, but its inefficiency has been previously presented. The trickle-down funds do not benefit all equally and prejudice low-income communities and the middle class. In a way, the defence of capital by the state could be justified by this aim of defending the raising of funds, but this assumption would require further revision.

This chapter offered a series of fundamental insights into the theoretical constructs of urban design under neoliberalism: A profit-oriented practice of urban design emerged not as a consequence of re-theorising consciously the discipline but as a reformation forced by the complete transformation of the Chilean society. The landmarks of this process are the transformation of the state ignited in 1975 after Friedman's letter to Pinochet, the National Policy of Urban Development of 1979, the Constitution of 1980, the democratisation of the country that started in 1990, and the deficit of more than a million housing units when democracy returned. This context forced a disciplinary change in urban design in order to provide urgent solutions to cities that lacked the time required for changing the existent institutional framework created by the dictatorship. The urban urgencies inherited from the dictatorship contributed to de-theorise and de-politicise the disciplinary field of urban design. A consequence of this accelerated and unreflective method of urban development is that free-market economics consolidated its position as the main guide for shaping the built environment. The market defined the aesthetics, criteria, and methods for producing cities, following the goals of profit and taking advantage of permissive regulations facilitated by authorities that were aligned with an economistic understanding of society. Urban design thus does not have a strong political body to defend its disciplinary transformation, and it succumbed to neoliberalism. It is necessary to unpack why this disciplinary body was not organised.

Finally, a refreshing finding has been the set of practices that are contesting urban design under neoliberalism. In my view, the one that may have the most impact is the new National Policy of Urban Development, but the practices presented by UKAMAU, MilM2, and Grupo TOMA also reveal some of the cracks in the virtual object. Thus, social engagement, grassroots processes of the production of spaces, and the public space as an opportunity for raising critique from the very activation of citizens are progressing towards the subversion of urban design under neoliberalism. Nevertheless, one of the elements that seems strongest and most difficult to tackle is the financialisation of housing. Embedded in the financial system, this feature of urban design under neoliberalism seems solidly constructed, and its transformation needs further information to arrive at an understanding of how to crack it. While this chapter has presented the current conflicts and conditions that have characterised Santiago under neoliberalism, the following chapter illustrates the practice of urban design under neoliberalism, with special focus on how urban designers face their everyday profession and what conflicts they have to face in the design of the city of Santiago.

Notes

1 Some of them were Pablo Barahona (President of Chilean Central Bank 1975–76 and Economics Minister 1976–78), Álvaro Bardón (President of Chilean Central Bank 1977–81; Sub-secretary of Economics 1982–83), Hernán Büchi (Economics Minister, 1979–80 and Finance Minister 1985–89), Jorge Cauas (Finance Minister 1974–77), Sergio de Castro (Economics Minister 1975–76; Finance Minister 1976–82), Miguel Kast (Planning Minister 1978–80; Labour and Social Security Minister 1981–82; President of Chilean Central Bank 1982), Felipe Lamarca (Director of Internal Revenue Service 1978–84), Rolf Lüders (Economics Minister 1982; Finance Minister 1982–83), Juan Carlos Méndez González (Budget Director 1975–81), and Juan Ariztía Matte (Superintendent of Private System of Pensions 1980–1989). These were the main advisors of Pinochet during the implementation of neoliberal policies, together with the lawyer Jaime Guzman.
2 This programme was developed under the name of El Ladrillo (The Brick), and it was previously presented to Jorge Alessandri for the election of 1970. At that time, Alessandri – being a conservative right-wing leader and former president – said that the programme was too radical for Chile (Fuentes & Valdebenito, 2015).
3 It is worth mentioning that the Chilean official motto is "by reason or by force".
4 Indeed, in 1997 the GG Publisher group launched a monograph about Klotz (one of many published in his honour).
5 In English: the problem of those with a roof.
6 In English: I like my neighbourhood.
7 Per capita GDP Chile 2016 USD 23,460 in purchasing power parity. Croatia was USD 23,596; and Greece was USD 26,383.
8 A shopping strip is a sort of small shopping mall that usually offers a small supermarket, a pharmacy, fast food, a café, and a parking area. In general, these urban products are located on one of the corners of a block.
9 The Hurun Report of 2017 indicates that real estate (9.4%) and construction (2.9%) are the sources of 12.3% of the wealth of billionaires in the world. The first source of wealth is technology and media at 13.2%.
10 Diverse specialists criticised the urban design of Costanera Centre. When a journalist asked Horst Paulmann about this criticism he said that Eiffel also was criticised for his Tower in Paris at the beginning but now everyone loves it (Cooperativa 2012).

5   The practice of urban design
in a neoliberal Santiago

Introduction

This chapter presents the ethical reflections of urban design practitioners under a neoliberal regime. The chapter focuses on the way urban design professionals see their own practice from a reflexive point of view and how they see themselves in the city they work in. It is important to recall that in Chile urban design, as a discipline and a practice, is conducted mainly by individuals trained as architects or planners. Therefore, in Santiago architects study a postgraduate course on urban projects (Master in Proyectos Urbanos at Universidad Catolica), urban development (Master in Urban Development at Universidad Catolica), or urbanism (Master in Urbanism at Universidad de Chile)[1] to deepen their understanding of urban design's disciplinary field. Alternatively, architects get a better understanding of the disciplinary field of urban design when working at public institutions (Municipalities, the Housing Ministry or the Public Works Ministry). The national Law of Urbanism and Construction (MINVU, 2017) provides that qualified architects are entitled to design buildings and act as urban advisors in public institutions (such as municipalities and ministries). Most of the interviewees are architects who work in the disciplinary field of urban design in Santiago.

The rule of profit in city making

The profit-oriented logic of neoliberalism implied a redefinition of the disciplinary field of urban design and particularly the processes of spatial production. Consequently, the mindset of urban designers shifted to endorse and adopt a neoliberal ethos. A whole new generation of Chilean architects and urban designers from the 1980s was educated with no consideration of the production of the urban as an independent matter from capitalism (Interviewee 15). Thus, urban designers find themselves at the centre of the implementation of the exploitative mechanisms of neoliberalism that operate through the constitution of a market-driven process of shaping the city. The urban process under neoliberalism was driven by the interests of capital, and the role of the urban designer in the process changed for the sake of accelerating the returns of rent from urban investments. As reported by Interviewee 15, before the beginning of the neoliberalisation of urban development the country had a smaller public budget to invest in public goods. However, even with less funding, social housing presented better qualities than after the return to democracy, when per capita GDP reached rates similar to those of Croatia or Portugal. The growth of GDP (usually employed for determining the economic success of a country) has no correspondence in the increase of the size of social housing units (Vergara & Boano, 2016).

Today, although Chile is a more prosperous country, social housing units are smaller than in the 1960s. This contradiction may have emerged from the methodology used by the state for selecting typologies of housing to build. As Interviewee 11 confirmed, profitability is the main criterion of allocation of public funds in urban projects:

> This is the way that city works: the Finance Ministry assigns resources to the Social Development Ministry (MIDESO) through the System for National Investments (SNI). The MIDESO evaluates projects and decides if they are recommendable or not. This recommendation is mainly based on the profitability of projects. This is an economic vision that contributes to increasing inequality because it tends to concentrate the investment in *comuna*s with higher populations or with better public goods that ensure the return of investments. Thus, the qualitative side of the social consequences of public investment is not considered, and effects such as segregation are reproduced. Innovation is zero. For instance, great urban improvements are considered as unnecessary. The cheapest projects have more chances to get funding. That's why urban design is not really considered in this formula. However, the richest municipality can afford *luxuries* such as good urban design.
>
> (Interviewee 11)

Profit has defined spatial transformations. This has happened in part because companies and entrepreneurs concentrated the tenancy of the capital to invest in the city but also because of state and urban policies that were transformed to accelerate the fluxes of capital, adopting profitability as the main criterion of action and thus redefining the methods used by public institutions, as well as private companies. This new entrepreneurial state (Hidalgo Dattwyler, Christian Voltaire, & Santana Rivas, 2017) influenced all levels of the urban governance system. This became more evident in the case of municipalities as they were left competing for resources: the richest municipalities concentrate more capacities and resources to develop good public spaces (Table 5.1). From the perspective of Interviewee 5, practitioners designing urban projects do not develop innovative projects because the revenues must be secured and the better way not risk a good profit is to use models that were successfully implemented in the city. Urban practitioners

Table 5.1 Incomes of municipalities and average income per household in each municipality. The spatial inequality between *comunas* may be illustrated by comparing the budget per capital of each municipality.

Municipality	Municipal budget per capita	Household incomes
VITACURA	774 CLP	4,444,561 CLP
LAS CONDES	762 CLP	3,474,634 CLP
PROVIDENCIA	652 CLP	3,133,694 CLP
LO BARNECHEA	599 CLP	4,591,334 CLP
SANTIAGO	330 CLP	1,190,959 CLP
HUECHURABA	279 CLP	1,092,563 CLP
LA REINA	191 CLP	2,660,650 CLP
QUILICURA	148 CLP	1,155,046 CLP
CERRILLOS	139 CLP	786,063 CLP
ÑUÑOA	137 CLP	2,114,814 CLP
SAN MIGUEL	125 CLP	1,650,387 CLP
RECOLETA	110 CLP	845,389 CLP
PEÑALOLÉN	109 CLP	1,451,987 CLP
MACUL	108 CLP	1,347,228 CLP
SAN JOAQUÍN	107 CLP	1,025,919 CLP
ESTACIÓN CENTRAL	106 CLP	995,784 CLP
INDEPENDENCIA	95 CLP	1,263,556 CLP
PUDAHUEL	87 CLP	1,063,766 CLP
LA CISTERNA	79 CLP	1,118,233 CLP
QUINTA NORMAL	79 CLP	956,333 CLP
RENCA	77 CLP	858,463 CLP
LA FLORIDA	62 CLP	1,211,707 CLP
CONCHALÍ	58 CLP	903,223 CLP
MAIPÚ	53 CLP	1,189,563 CLP
PEDRO AGUIRRE CERDA	30 CLP	889,094 CLP
SAN RAMÓN	28 CLP	828,719 CLP
LO PRADO	26 CLP	873,375 CLP
EL BOSQUE	25 CLP	836,300 CLP
LA GRANJA	23 CLP	993,292 CLP
LO ESPEJO	21 CLP	755,101 CLP
CERRO NAVIA	18 CLP	842,726 CLP
LA PINTANA	17 CLP	830,330 CLP

Source: Own elaboration based on CASEN 2015 and Sistema de Información Municipal.

have normalised the way that public institutions deal with transformations in the city based on criteria related to financial evaluations and positivist understanding of spatial problems, just like the IRR and NPV under the scheme of cost-benefit approach for evaluating urban projects. In the development of new projects, urban designers "must find a source for funding it in the government. . . . The problem is that everything is governed by specific regulations for defining investments; there is no innovation in public projects. The state does not innovate. Perhaps in some specific cases of municipalities with great budget some mayors dare to innovate, but never from central government. It is not usual at all" (Interviewee 5).

In this scenario of a constrained capacity for innovation, urban designers must accept the limitations imposed on their creativeness, adopting economic creativity as new skill. For example, in the case of social housing, the Housing Ministry has a set of predefined designs, standards, and spatial norms that architects may apply to build projects in specific communities. Indeed, using these predefined designs is convenient because the samples are pre-approved by the ministry, and the process of construction may therefore begin earlier. In general, these designs are quite similar: made of concrete blocks, a maximum height of five storeys (generally four storeys), no elevators (too costly and too time-consuming to build), composed of flats with two rooms, one bathroom, kitchen, and a common area. Typical apartments have an average area of 36 m^2 to 44 m^2. Recently, the government has included new and better designs, but most social housing projects comply with these features. Urban design under neoliberalism produced an inhibition to innovate and therefore a homogenised landscape in relation to social housing but also in real estate development. Both typologies follow the same criteria, and the main difference is that real estate has a bigger budget. The conflict for urban designers is that projects do not have a main objective that values the imagining of a better future for communities and cities at large; the goal is, put simply, to provide "ready made" and codified products ensuring that investment generates acceptable rates of profit while keeping down the costs of production (time of production and design solutions). As Interviewee 9 explained, "when working with public policies, you must recognise what are the real possibilities for new developments. This is a bit frustrating because, in my experience, the most interesting designs or, if you prefer, the more radical proposals for urban projects cannot be implemented in the end. The capacity of urban designers' lobbying is vital".

Under an extremely deliberative scheme, urban design enters a phase of irrelevance because spatial proposals going beyond the limits of the common ground are immediately excluded from the possibilities of being even discussed. Also, the common ground, the discussions for finding agreements, or the feasibility of urban projects in a neoliberal context depend mostly on the interests of the capitalists. In this scenario, where diverse forces struggle and compete in shaping the city present and future, one expects to see an unbalanced decision environment in which the city becomes a representation of the interests of capital. The case of the CChC emerged constantly in this research as it has the capital and the organisational structures to strongly influence a very specific idea of urban development to prevail. Over the decades, the CChC has employed all its influential power to foster property rights, and its success is evident. Nowadays, in Chile, the main priority of people is homeownership (Figure 5.1) and the main pathway to its achievement is through personal efforts and meritocracy (Mayol, Azocar, & Azocar, 2012). This is also reflected in a decline of collective discussions about the future of cities, their public nature, structure, and places. The city as a collective product is at stake because increasing individualism has characterised the country where eight out of ten distrust people different from their families (COES, 2015). Considering the individualistic stance that emerged from neoliberalism, dissent is avoided and preference is given to a trench-on strategy of fighting ideas without entering into confrontation. With this logic, it is to be expected that the stronger party wins the battle. Interviewee 2 illustrates this issue:

> I see a lot of ingenuity by the state when searching for social justice. There is an excessive confidence in capital, but when housing and

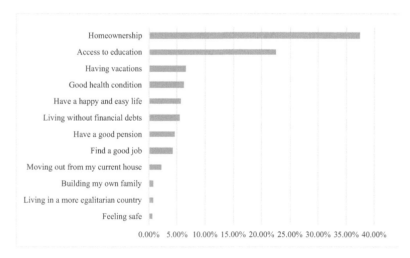

Figure 5.1 Chart that summarises the wishes and dreams of Chileans.

Source: Own elaboration based on the study "Chile Dice" elaborated by Universidad Alberto Hurtado in 2017.

transport tend to become oligopolies, it is logical that the market will not be efficient in developing self-regulation. Asymmetries emerge due to the uneven distribution of good and services in the city

(Interviewee 2)

Under neoliberalism, each social actor attempts to defend its positions and capitalists avoid any risk of jeopardy to the efficiency of capital investments in the city. Thus, when the state adopts the methods and strategies of entrepreneurialism – represented in space as urban design under neoliberalism practices – it starts to foster the fluid circulation of capital and enhances its performance. As mentioned in the previous chapter, there is an excessive care paid to the internal rate of return and net present value as the measures to define an urban project. This trend is particularly visible in the segregation of housing projects showed in Figure 5.2: buildings are similar but in Las Condes each square metre has £1,000 more value invested than do other *comunas*.

Poorer *comunas* have smaller dwelling space than richer *comunas*. Also, it is visible that household incomes are related to housing price. As Milton Friedman (1975) claimed, the reduction of barriers to the flow of investments was assumed by Chilean institutions, especially in relation to urban policy.

Following the interviewees' affirmations, it seems to be a fact that the city has been shaped following the expectations and plans of companies, mostly oriented to better allocating investments in order to obtain short-term revenues from urban development projects. In this entrepreneurial scheme, the state has provided funds only for investing in specific areas of the city, resisting involvement in any other roles. The inequalities that are reproduced by the redistributive weakness of local institutions are aggravated by the general dismantling of the regulatory capacities of public agencies and reinforcement of market discipline.

State, neoliberalism, and urban design

While examining how the municipal planning departments and the ministries manage urban processes, it is necessary to reflect on how the image and the vision of the city are constructed by the regulatory framework and what market forces attempt to conduct the urban process to extracting value from cities. Interviewee 6 has vast experience working in public institutions and provided a reflection of how the state is apparently not much involved – or interested – in defining the image of the city:

In the Architecture Department of the Public Works Ministry we do not design spaces but only execute public policy. In fact, the Ministry does

Figure 5.2 Sample of real estate projects in 4 different *comunas* in Santiago.
Source: Own elaboration.

not design public policies either; they come defined from the presidency or the Congress. This represents a great fragility because there is no diagnosis of urban issues from the state. Instead, we simply trust in the projects developed by the central government or some private company. With private companies, if they present a valuable project for the city that meets the feasibility criterion, we sign an agreement for its implementation, and that's it. Therefore, in many cases, the private sector executes the projects using the state as investor. No risks at all

(Interviewee 6)

What emerges from the interview is that the state's involvement in the development and the design of urban vision and form is quite limited not only because of limits to the availability of funds but also because aesthetic reflections are not the state's responsibility. In seeking to unveil the reasons behind the abandonment of aesthetics discussions, some interviewees elaborated that the cause was that public institutions lack a long-term vision for cities and thus the imagining of the built environment is ignored. This is a consequence of the fact that decisions are mostly in the hands of politicians whose commitment depends on how long they will be in administration.

For instance, some interviewees mentioned that mayors still make most of the decisions about urban spaces, particularly public spaces and urban facilities. As reported specifically by Interviewee 12, who works as advisor in a municipality:

My practice as urban advisor for the municipality is developing and monitoring the Communal Regulatory Plan. In order to do so, I simply follow the methodology that is very commonly known: research urban changes in the district, and review demographic trends, variations in the land market and the needs of the community, all based on what is stipulated in the Ley General de Urbanismo y Construcciones. In Chile, nowadays, urban planning is only about regulation. The state cannot actually decide how to invest because it is used to regulating only.

(Interviewee 12)

The instrument known as the Plan Regulador Comunal is the principal planning instrument used by Chilean municipalities to define the urban image. However, it is quite limited because it works only by zoning areas of the city for specific functions (industrial, commercial, residential, leisure). Considering that the law's function is only regulatory, the capacity of the municipalities for deciding where to invest seems absent.

It is important to make the point that in Chile municipalities must be self-funded. It means that each municipality works as an enterprise that needs to ensure its yearly income. All municipalities in Chile receive a basic income provided by the Fondo Comun Municipal (Municipal

Common Fund) that allows the payment of rents and salaries for work-ers and some basic functions. All other investment emerges from the col-lection of municipal patents and taxes. Of course, this is a method that perpetuates the uneven distribution of public funds. For instance, the most expensive properties are located in the wealthy *comunas*. The territorial tax that these properties pay is high, so the municipality receives more funds than lower-taxed districts. This is completely the opposite in low-income *comunas*, where the properties are cheaper and residents pay lower taxes, leaving municipalities short of funds to invest in public spaces and facili-ties. At a national level, the concentration of incomes of municipalities per taxes are as follows: Las Condes 9%, Santiago 6%, Providencia 4%, Lo Barnechea 4%, Vitacura 3%, which is correlated with the concentration of wealthy people and availability of parks and green areas as a share of the region (Table 5.2). Of course, the wealthiest people in the country are concentrated in these *comunas*.

Most of the investments in public spaces funded with public funds need to demonstrate their profitability based on the methodology developed by Social Development Ministry (MIDESO). The National System of Invest-ments (SNI) determines the social prices of public investments (Sistema Nacional de Inversiones, 2017). For public spaces, SNI demands the elabo-ration of a project profile that fulfils the requirements of the cost-benefit approach, which must (i) identify benefits of investments, (ii) quantify those benefits, (iii) identify costs, (iv) quantify the costs, and (v) indicate profit of investment and elaborate the indicators of cost-efficiency of the project (MIDSEO, 2013).

The major conflict when designing public spaces is the scarce avail-ability of financial resources. This limits the expectations of people about their cities. When you work under a scarcity of resources, you

Table 5.2 Distribution of income per taxes, concentration of wealthy people, and green areas per inhabitant.

Comuna	Income per taxes of municipality over the national revenues	Concentration of wealthy people over the national level. (A, B1, and B2 income groups)	Green areas per habitant over the metropolitan region
Las Condes	9%	23.2%	10.82%
Santiago	6%	1.2%	8.59%
Providencia	4%	17.8%	7.28%
Lo Barnechea	4%	3.6%	4.04%
Vitacura	3%	4.1%	6.2%

Sources: own elaboration based on Asociacion de Municipalidades, CASEN 2015.

Table 5.3 Rank of availability of municipal budget per capita in Santiago, compared with household incomes and green areas per capita.

Rank	Municipality	Municipal budget per capita (£)	Household incomes (£)	Green areas per capita (m²)
1	VITACURA	0.92	5,291	16.74
2	LAS CONDES	0.91	4,136	9.07
3	PROVIDENCIA	0.78	3,731	11.67
4	LO BARNECHEA	0.71	5,466	9.08
5	SANTIAGO	0.39	1,418	5.52
6	HUECHURABA	0.33	1,301	3.67
7	LA REINA	0.23	3,167	10.99
8	QUILICURA	0.18	1,375	2.89
9	CERRILLOS	0.17	936	6.14
10	ÑUÑOA	0.16	2,518	3.96
11	SAN MIGUEL	0.15	1,965	1.75
12	RECOLETA	0.13	1,006	2.15
13	PEÑALOLÉN	0.13	1,729	3.53
14	MACUL	0.13	1,604	3.59
15	SAN JOAQUÍN	0.13	1,221	2.15
16	ESTACIÓN CENTRAL	0.13	1,185	3.7
17	INDEPENDENCIA	0.11	1,504	0.87
18	PUDAHUEL	0.1	1,266	1.37
19	LA CISTERNA	0.09	1,331	1.22
20	QUINTA NORMAL	0.09	1,138	1
21	RENCA	0.09	1,022	1.99
22	LA FLORIDA	0.07	1,443	3.04
23	CONCHALÍ	0.07	1,075	2.44
24	MAIPÚ	0.06	1,416	4.23
25	PEDRO AGUIRRE CERDA	0.04	1,058	0.99
26	SAN RAMÓN	0.03	987	2.8
27	LO PRADO	0.03	1,040	1.78
28	EL BOSQUE	0.03	996	1.48
29	LA GRANJA	0.03	1,182	1.89
30	LO ESPEJO	0.02	899	1.3
31	CERRO NAVIA	0.02	1,003	2.49
32	LA PINTANA	0.02	988	3.07

Source: Own elaboration based on Asociacion de Municiaplidades.

must maximise the investment doing durable projects, and those solutions are not necessarily interesting.

(Interviewee 6)

In the case of Santiago, an urban ensemble of 32 municipalities, it is difficult to ensure that public space will be distributed equally maintaining the same quality. The budget varies depending on the income of the municipality, as mentioned; therefore, access to public spaces and facilities and their quality depend on the socio-economic characteristics of the municipal community: the richer and the wealthier, the better equipped. Given that the access to high-quality public space is not for everyone in the sense that it is not conceived as a collective urban dimension, the city becomes discriminatory, developing exclusionary mechanisms and unequal access to good spaces.

> Public goods should be a luxury that all citizens have access to. This luxury should be a right materialised in the public space. The access to public space as a luxury is a responsibility of the state, but it is not ensured for everyone. It is visible when walking through Lyon Avenue [Figure 5.3] in Providencia and then you go to Bajos de Mena [Figure 5.4] or La Pintana. It is not democratic, and it should be.
>
> (Interviewee 26)

The urban designer attempts to develop good designs, but the asymmetries of resources between municipalities undermine the provision of more even

Figure 5.3 Ricardo Lyon Avenue in Providencia.
Source: Own elaboration.

Figure 5.4 Santa Rosa Street in Bajos de Mena.
Source: Own elaboration.

results between different areas of the city. Spatially, this unevenness of public facilities is a symptom of segregation. As municipal incomes depend on
how wealthy a district's residents are and the consumption by low-income
population of public goods is concentrated in certain *comunas*, urban life
deteriorates in part of these *comunas* while urban life is enhanced in high-
income areas, reproducing the problem.

This condition of unevenness is structural; however, urban designers
have not been particularly active in contesting these conflicts, and advo-
cacy for a more just city is not organised. "There are no structural advances
from the urban development model implemented by the dictatorship and the
current model. There are some add-ins but not real changes. It is difficult to
say why, I do not have a straightforward answer" (Interviewee 1). Subsequently, "in Chile, there is no urban planning, it is extinct, we only have the
market that defines the city form" (Interviewee 15). The extinction of urban
planning comes from the state apparatus. Given this context, it is important
to identify how the market facing the city is making progress in order to
understand how urban design under neoliberalism actually operates.

Profit-oriented urban design

Real estate developers still have plans for the city, although these plans are not necessarily an image or a blueprint; instead, the plans may be better represented in numbers. These plans are mostly oriented to defining strategic investments for extracting value from urban products. As the law compels, these plans are supported by architects and urban designers. So, what is their stance in relation to producing a better city? It is important to know how urban designers working for private firms are facing the critiques coming from colleagues in academia and state.

> It bothers me when colleagues blame real estate companies while urban designers are allowing these kinds of projects. It may be corruption or influence peddling as well, but in the end, there is always an urban designer signing a project because in Chile this is obligatory by law. Regulatory plans, projects and other urban ideas must be signed by a certified urban practitioner. What is real is that real estate companies hire urban designers and they must do what their bosses ask. As every urban designer wants to preserve their job, the city's development depends on unprincipled ideas. That is why nowadays real estate companies are out of control too.
>
> (Interviewee 16)

> The city has become a derivation of the economic model, a direct effect. The grim symbiosis between the urban model and the economic model has been accepted. It seems like people believe blindly in free entrepreneurship that has transformed urban discussions into ideological issues. These groups have ideologised urban development through a frenetic defence of neoliberalism that was imposed by force. The Chilean neoliberal experiment is directly related to crimes, and this contributes to an ideologised discussion. Also, this has strengthened the resistance to solidarity and to imagining a better city. I believe that this is because the Chilean elite is ignorant, especially the entrepreneurial elite.
>
> (Interviewee 26)

Based on the historical role on urban processes of the Chilean elite, for them it is important to use methods capable of producing good spatial results although always subjugated to how these results may ensure high profits. It seems that urban professionals have not been efficient enough to change and contribute to a public debate on the value and form of the city as well as able to lobby the elite by showing different and better means to producing cities for good and not only for profit.

The dictatorship and the forceful implementation of neoliberalism succeeded in changing the understanding of the practice of urban design in a short period – not only by fostering the profit-oriented logic in the development of the city but also by transforming the way that practitioners relate to citizens and the practice of building the city.

As is shown in Figure 5.5, between 2010 and 2016 the process of building Santiago was directed by the profitability of investments. The tables and maps of Figure 5.5 are self-explanatory: wealthy *comunas* – such as Las Condes, Santiago, Providencia, Lo Barnechea, and Vitacura – are always in the list of the 15 *comunas* that allocate most of the investment in relation to real estate development projects. Although it is difficult to demonstrate coordinated action by building companies in deciding where to invest, the patterns and the trends are clear. Also, as there are some *comunas* where investment in real estate projects is fixed, some low-income *comunas* are also subject to a good amount of investment depending on the year. It seems that real estate capital is allocated for secure investments in high-income *comunas* while it is used for speculation in lower-income *comunas* to test the ground for further investments in the future. If the investment in low-income *comunas* is profitable enough, then the companies start to invest higher amounts of capital in these *comunas*, and more companies participate in the market of each specific area. If investments are not profitable, then they move away to another *comuna*.

From the interviews it emerged that urban designers working for the state have quite limited capacities for designing the city. Instead, these professionals almost exclusively spend their time checking that the private contractors fulfil certain criteria for shaping the space. Therefore, for urban designers working for public entities, their imagination is not required. It is in the private sphere that it is still possible to find urban designers shaping spaces, although they also function under the dogma of profit.

At least in Santiago, urban designers working in the private sector still depend on the investors and their expectations of profit. When you are an employee, it is difficult to develop some innovative project, especially when the status quo is particularly profitable for the people who hired you. For example, one of my students was working in an architecture studio, and he saw the possibility of improving the design of a residential building's windows. He proposed the idea to his bosses, but it was rejected. One of his bosses, who also lectured him in the School

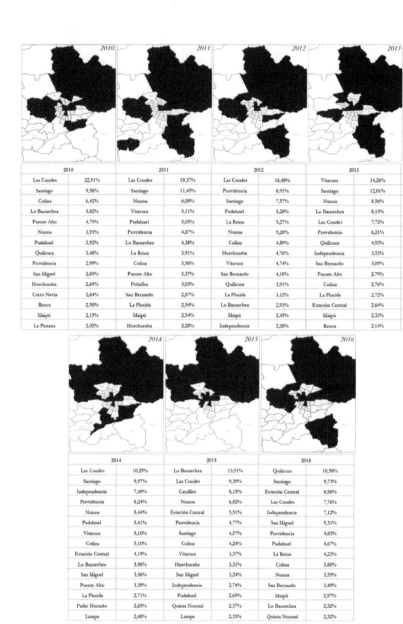

2010		2011		2012		2013	
Las Condes	22,91%	Las Condes	19,37%	Las Condes	16,48%	Vitacura	14,26%
Santiago	9,58%	Santiago	11,45%	Providencia	8,91%	Santiago	12,01%
Colina	6,42%	Nuñoa	6,09%	Santiago	7,57%	Nuñoa	8,56%
Lo Barnechea	5,82%	Vitacura	5,11%	Pudahuel	5,28%	Lo Barnechea	8,15%
Puente Alto	4,79%	Pudahuel	5,05%	La Reina	5,27%	Las Condes	7,72%
Nuñoa	3,93%	Providencia	4,87%	Nuñoa	5,20%	Providencia	6,21%
Pudahuel	3,92%	Lo Barnechea	4,38%	Colina	4,89%	Quilicura	4,93%
Quilicura	3,48%	La Reina	3,91%	Huechuraba	4,76%	Independencia	3,53%
Providencia	2,99%	Colina	3,56%	Vitacura	4,74%	San Bernardo	3,09%
San Miguel	2,85%	Puente Alto	3,37%	San Bernardo	4,18%	Puente Alto	2,79%
Huechuraba	2,69%	Peñaflor	3,03%	Quilicura	3,91%	Colina	2,76%
Cerro Navia	2,64%	San Bernardo	2,87%	La Florida	3,12%	La Florida	2,72%
Renca	2,58%	La Florida	2,54%	Lo Barnechea	2,93%	Estación Central	2,64%
Maipú	2,15%	Maipú	2,54%	Maipú	2,45%	Maipú	2,33%
La Pintana	2,02%	Huechuraba	2,28%	Independencia	2,28%	Renca	2,14%

2014		2015		2016	
Las Condes	10,29%	Lo Barnechea	13,91%	Quilicura	10,58%
Santiago	9,97%	Las Condes	9,39%	Santiago	9,73%
Independencia	7,49%	Cerrillos	8,15%	Estación Central	8,96%
Providencia	6,24%	Nuñoa	6,82%	Las Condes	7,76%
Nuñoa	5,44%	Estación Central	5,51%	Independencia	7,12%
Pudahuel	5,41%	Providencia	4,77%	San Miguel	5,31%
Vitacura	5,10%	Santiago	4,57%	Providencia	4,83%
Colina	5,10%	Colina	4,24%	Pudahuel	4,67%
Estación Central	4,19%	Vitacura	3,37%	La Reina	4,23%
Lo Barnechea	3,96%	Huechuraba	3,31%	Colina	3,80%
San Miguel	3,56%	San Miguel	3,24%	Nuñoa	3,55%
Puente Alto	3,39%	Independencia	2,74%	San Bernardo	3,49%
La Florida	2,71%	Pudahuel	2,69%	Maipú	2,97%
Padre Hurtado	2,65%	Quinta Normal	2,37%	Lo Barnechea	2,32%
Lampa	2,48%	Lampa	2,33%	Quinta Normal	2,32%

Figure 5.5 The figure indicates the percentage of square metres constructed per *comuna* of the total square metres built in Santiago per year between 2010 and 2016. The maps reveal the way that the city has a clear tendency to concentrate investment on the east side of the city.

Source: Own elaboration based on INE data.

of Architecture, told him: why do you bother so much if you are not going to live there? This was not the first time that I heard this kind of story, and I believe that this clarifies a bit the ethical distortion in our colleagues.

(Interviewee 4)

This quotation from Interviewee 4 is disturbing not only because it is a reference to a professional but also because it makes reference to a lecturer in urban design talking to a student. The role of urban designers in the neoliberal context of Santiago is mostly valued from an economic perspective, how profitable, efficient, and fast a proposal can be. This is frustrating from the perspective of the artistic creativity that urban designers should have (Barnett, 1982; Matthew Carmona, 2014b). While urban design usually has a political, cultural, economic, and social role (Cuthbert, 2011), in Santiago this seems not to be the case. The practice of urban design seems to have abandoned any of those roles and is instead relegated to a simple method for the mere execution of norms for producing space. As the interviews advanced, I noted a worrying question: Is it possible to argue that urban design under neoliberalism eliminated urban design? Perhaps the pursuit of designing good cities for all was replaced by a new ethos of designing cities for profit.

> In Chile, there are no urban designers but architects who are specialised and in the end they facilitate land accumulation for the elite. There is an evident cultural and educative influence of architecture on urban development in Santiago. The city is a sum of objects rather than a coordinated effort to comprehensively organise functions in space.
>
> (Interviewee 6)

> In Chile, there are no urban designers; there is no diagnosis before developing a project. What we have in Chile are market planners. They only follow the rules of the market. Some public constraints make urban discipline in Chile difficult, but in the end, the diagnosis comes from outside of the urban
>
> (Interviewee 2)

The difference between an urban designer and a marketing manager[2] in Santiago seems to be marked by a thin line. The marketing manager is a professional – generally an economist – whose specialty is defining strategies for selling any kind of products in certain segments of the population, defining prices based on these segments, and also creating new demands for new products that ensure profits. In the end, both specialists – urban designers and marketing managers – develop their practice basing decisions on the information about the market, and both try to accomplish the expectations of consumers while ensuring high revenues for investors.

Challenging urban design practice in Santiago

> Urban designers are unnecessary thinkers. We are only technicians who execute projects. Thus, we have two options: either you become a technician working for the state or some consultancy, or you become marginal
>
> (Interviewee 7)

> Urban designers are over-ideologised or vulgarised. They do not comprehend the processes of investments, and when the city appears on the political agenda, it is not due to urban designers, so they only react to exogenous requests
>
> (Interviewee 11)

Criticism about the lack of political training that prevents urban designers from becoming actual agents of change in Santiago illustrates one of the main challenges facing urban design under neoliberalism: theorise its praxis. By operating as spatial specialists and leading processes of production of the spaces, urban designers have an advantaged position for pushing political agendas that may politicise spatial transformations and infuse society with an understanding of the importance of controlling the future of their cities. From the interviews, it is possible to argue that the absence of discussions and projects imagining the far future of Santiago stems from the way that urban designers were absorbed by neoliberalism and did not question their role in the circuit of capital.

> The city has been fragmented into pieces by each discipline, with a huge role played by economics. This is represented in a political structure organised by investment targets depending on the magnitude of projects. For example, the Public Works Ministry is considered as a contributor to economic development in Chile because it invests in infrastructure, roads, ports, and facilities for capital flow. On the other hand, the Housing Ministry is considered as a cost for the national budget because it provides housing and works for society with a focus on low-income

communities. In the end, this fragmentation eliminates the city from the discourse

> (Interviewee 5)

The fragmentation of the city has produced urban designers who are disconnected, and thus the possibilities of thinking the city collectively have been diminished. This observation criticises the fragmentation of urban institutions, which influenced the way urban designers worked and developed projects. It also divided various practitioners, so their possibilities for engaging in a broader claim for changing their practice seem hampered by a fragmented reality. This could be a situation of an alienated body of professionals whose goals are similar (shaping the space) but whose cohesion is low because the system in which they work is designed to avoid the organisation of collective views of their practice. As a result, the future idea of Santiago is the result not of a collective reflection but only of a series of isolated ideas for the city.

> Nobody imagines Santiago in the future. We are not even dreaming it because it must be a collective construction. It cannot emerge from the individualistic society in which we live in now. Who is doing a collective discussion of the city nowadays? Nobody, nobody is doing so.
>
> (Interviewee 15)

It is important to mention that diverse groups in society have come forward to break the alienation and think collectively about a new future for Chilean society. Particularly in the case of education and social security rights, diverse social movements have occupied cities and paralysed the country more than once since 2011. It is the awakening of the Chilean society from the neoliberal dream (Castells, 2015) or nightmare à la Aldous Huxley in *Brave New World*. Diverse authors have spoken of the importance of a new generation of politically active citizens whose idea of democracy is not being fulfilled by Chile's political structure (Ruiz & Boccardo, 2014; Salazar, 2012b; Valenzuela, Penaglia, & Basaure, 2016). In this context, the interviewees reflected on the significant opportunity that these movements represent for reactivating the importance of urban design in Santiago.

> A democratic city is being organised in the streets. Today it is impossible to modify a regulatory plan without consulting the citizens.

> Communities are rising when they want to defend their neighbourhoods. This practice is getting more common.
>
> (Interviewee 9)

> The only way to recover urban design for people is through popular mobilisation. It is difficult, but organisation is needed.
>
> (Interviewee 15)

Interviewee 18 reflected on the importance of developing a different intelligence capable of uniting people to defend the right to the city. To do so, it is necessary to highlight a claim, an urban contradiction that affects the life of thousands and that can trigger a broader demand from citizens for deep transformations. This contradiction must be visible to urban practitioners, just as students explained how education was biased in Chile. One of the agreed claims that may unite masses of people in struggling for urban design is housing.

For Alfredo Rodriguez and Paula Rodriguez, the poor quality of social housing is representative of the neoliberal city in which, rather than a policy of social housing, there is a policy of cheap housing solutions. In their view, the housing solutions constructed in Santiago are a social problem that goes straight to the right to the city of marginalised people living in these new *comunas* (Rodriguez & Rodriguez, 2009). Diverse authors who have studied the neoliberalisation of Santiago tend to focus on housing because in the end it represents the most basic space to which people struggle to get access.

> The excess we see today is inconceivable. Housing as a commodity is the worst. It is impressive the number of housing blocks built with awful standards. Many must be demolished in Bajos de Mena, Rancagua, Temuco and so on. Visiting these districts is like visiting bombing zones in Gaza. People coexist with some of these buildings that the state still is attempting to fix while others have already been demolished. You can see the skeletons of these buildings scattered in cities.
>
> (Interviewee 15)

The commodification of housing represented a means for igniting cycles of capital accumulation from construction activities but also squeezing household budgets, disregarding an already diminished income despite the high costs of life in Santiago. For instance, the average return on buildings in the city of Santiago during one recent seven-year period was an internal rate of return of 47%, while in London it was around 20% and in Manhattan 24%. If an IRR of greater than 20% is highly profitable, the IRR of 47% provides evidence to argue that in Santiago the construction business is an outstandingly profitable activity. This criterion also applies to projects funded with tax revenues. For example, the average profitability in the construction of roads in cities is 44% on average, as shown in Table 5.4.

> Santiago is an uneven metropolis as a totality: it is totally unequal, with uneven spaces, variable speeds, different materiality, polarisation and contrasts. Santiago is the geography of unevenness, with a rich top and a poor bottom, a non-levelled playing surface with a clear leaning.
>
> (Interviewee 3)

The unevenness of Santiago is quite well represented in Table 5.5, which illustrates the costs of living in Santiago depending on one's socio-economic quintile.

> The way that politicians provide solutions to cities is unbalanced. Citizens do not receive benefits for living in dense and compact cities, but instead, they have to pay the costs of the improvements: high prices of tickets for transport, expensive basic goods, most of the salary goes to pay accommodation, education, and food.
>
> (Interviewee 8)

The invisible hand of the market resulted in an apparent condition of spatial inequality. This inequality has been quantified by analysing the segregation of Santiago. For example, the Dissimilarity Index (for measuring residential segregation) in Santiago is 51%, which is high in comparison to that for Amsterdam or Athens (van Ham, Tammaru, de Vuijst, & Zwiers, 2016), as shown in Table 5.6.

> The extreme Chilean neoliberalism has made of Santiago a principal city for Latin America. It is an extremely efficient business centre and a safe reservoir for capital. Almost like a potential state-city of global capitalism. Santiago is a global neoliberalism reserve. It seems as if

Table 5.4 Profitability rates of projects funded with tax revenues.

	Project 1: Nursery in La Cisterna	Project 2: Football court Buin	Project 3: Road maintenance in metropolitan region	Project 4: Surgery in La Florida	Average
IRR	**23%**	**55%**	**49%**	**49%**	**44%**

Source: Own elaboration based on information of Mercado Publico.

Table 5.5 Cost of living for a family living in Santiago by socio-economic quintile.

Quintile	I	II	III	IV	V	Source
Average income	£318	£375	£474	£662	£1.958	*Based on CASEN 2015 (yautcor)*
Renting	£194	£214	£245	£317	£570	*Based on CASEN 2015 (v22)*
Basic goods	£162	£188	£207	£219	£205	*Based on Encuesta de Presupuestos Familiares*
Transport	£120	£139	£153	£162	£152	*Transports Ministry considering 2 daily rides for 20 days*
Balance	***-£159***	***-£166***	***-£131***	***-£35***	***£1.030***	
People per house	3.37	3.9	4.29	4.54	4.27	*Based on CASEN 2015*
% income spent on rent	**61%**	**57%**	**52%**	**48%**	**29%**	*Based on CASEN 2015*

Source: Own elaboration based on CASEN2015, EPF2013, INE and Transport Ministry.

Table 5.6 Dissimilarity Index of Santiago compared with that for sample European cities.

Santiago	Madrid	London	Stockholm	Vienna	Amsterdam	Athens
51%	49%	42%	40%	40,50%	33%	35%

Source: Own elaboration and van Ham et al. (2016).

> the branding of Santiago attempts to transform this city into the most neoliberal one in the world
>
> (Interviewee 21)

In an era when cities are branded and marketed, the potentialities of Santiago as the most neoliberal city in the world seem like an interesting reflection. Many of the interviewees observed the problematic positivist and profit-oriented logic of the system for defining public investments, which undermines the possibilities of developing a city in which urban life is accessible for everyone. Interviewee 12 demands a change:

> A key reform is changing the methodology of public investment used by the Ministry of Social Development (MIDESO). The

subsidiary bias is clear. Subsidies for housing and public transport are delivered without coordination, which undermines the development of a good city

(Interviewee 12)

The effort to adopt this methodology for evaluating urban development comes from a desire to ensure that public funds are employed wisely. "The method for evaluating a public project investment hampers the possibility of developing multidisciplinary approaches. Indeed, there is a structural disincentive to cooperate, fostering fragmentation between public agencies" (Interviewee 2). Quantitative methods are more respectable under a neoliberal scheme.

Public policies aim to invest with exactitude. That's why in Chile the mathematic model is so used for evaluating investments. The goal is optimising resources. Since 1973 the state has been dealing with the same model. That is why Aylwin massified the production of housing, privileging quantity over quality. As a consequence, a colossal segregation was created which in the end represented a more significant problem than the housing deficit.

(Interviewee 16)

Political commitment is a key challenge for urban design under neoliberalism. Five out of 27 interviewees mentioned the issues of apathy and a lack of political interest among urban designers, meaning they lack the actual capacity for influencing the way the city is produced. These interviewees suggested the urgency of moving towards more politically informed urban practices – towards the politisation of urban design practitioners – to increase the importance of the city in the political arena, engaging in the defence of urban life, and promoting the idea that the city is fundamental for social justice.

Transformed into a strategy for profit, urban design under neoliberalism has prioritised specific modes of thinking the city. None of these methods has advanced decidedly towards a more collective conception of the city, nor has any of them engaged practitioners for redefining their ethos, and individualised practices reflect on the society as a whole.

People who were called to develop a utopian idea of the city never did it. Now, the scale of Santiago makes it impossible to think of a new image of the city. The return to democracy strengthened speculation with land and the massive production of housing. . . . Afterwards, urban development will be associated with these housing towers, and the urban designers will be only spectators of this history.

(Interviewee 11)

The aesthetics of urban design under neoliberalism seem to be represented by the set of high-rise towers for housing constructed in Santiago's downtown and other *comunas* (Figure 5.6) – a pantone of colours and materials that have become monotonous in the urban landscape of Santiago.

> Santiago lost the capacity for imagining itself. For instance, the Metropolitan Regulatory Plan of Santiago (number 100) only set changes in the political-administrative limits of the urban area. Imagining the city is not even mentioned, and of course, utopia is not taken into consideration. There are no more utopias. We are trapped by an image of the city as an efficient machine for producing wealth.

(Interviewee 2)

The majority of the interviewees had a more pessimistic idea about the future of Santiago, mainly because of the lack of social engagement in relation to urban affairs and urban designers' relatively low capacity to influence discourses and policies.

> The prospects of a future Santiago were developed in the past, around the fifties. Today there is no much reflection about it. We must imagine Santiago for the future because neglecting it is blindness for public policy. . . . We, as urban designers, are not in a propositive stance but instead, we are reactionary.

(Interviewee 9)

> Santiago nowadays is going directly to the cliff, but we are attempting to stop it from the streets. Some initiatives may refer to some utopian thinking that has emerged from academia and social roots.

(Interviewee 2)

As a final reflection on these challenges, it seems that the effort of those urban designers aiming to save Santiago from its total neoliberalisation will rely on a practice embedded in social movements – encouraging people to contest urban design under neoliberalism in the streets, marching and protesting – and also require a comprehensive and collectively produced political agenda for subverting neoliberalism. By taking this path of contesting the current situation of their practice, urban designers seem to constitute

Figure 5.6 Set of housing towers built in Santiago's downtown as a collage of textures and aesthetic proposals of real estate development in Santiago since 1990.

Source: Own elaboration.

a potential resource, from different practices and approaches, for reconfiguring their practice and politicising ways of understanding the production of spaces that call for what may be seen as urban solidarity.

Concluding remarks

This chapter explored the approaches and attitudes of urban design professionals to their practice in Santiago, enquiring how they observe and criticise the ethos of their disciplinary field colleagues under a neoliberal regime. The chapter served to illustrate forms and values of how professionals of space in the neoliberal city of Santiago understand the problems related to urban design. As a result, this chapter provides evidence for elaborating a reflective critique in the attempt to explain urban design under neoliberalism and its socio-spatial consequences for constructing meaning for urban design under neoliberalism from the perspective of everyday practices.

A concerning finding is that urban designers in Santiago recognise their political irrelevance, despite acknowledging that urban design is quite important for advancing towards a just society. Furthermore, the artistic, imaginative, and innovative importance of design is missing from the interviews I conducted. An aesthetic reflection on neoliberal spaces is not part of everyday urban design in Santiago. Urban design under neoliberalism is under-theorised, lacking epistemological reflections about the ethos of this disciplinary field. My reflection is that the excessive importance given to private property and the rapid privatisation of urban life fostered by the neoliberal ideology explain – in part – why urban design and public spaces are not much considered when talking about urban specialists in Santiago. This allows me to state that the future of public space in Santiago under neoliberalism is conflictual and may face privatisation if it is to survive. From these reflections, it is possible to consider that the city of Santiago since 1979 has been produced and constructed simply through the aggregation of private plots whose articulation of the public space is weak. Therefore, during past decades, the development of public spaces in Santiago has not been a priority. In this scenario, the imagining of good spaces and the pursuit of a better city do not fit in; they do not suit the scheme. This contradiction may be the trigger for an emerging disciplinary transformation to revolutionise the practice of urban design in Santiago. It would require a different ethical definition and also a theory for practice.

Notes

1 The National Agency of Postgraduate Programmes of Chile recognises only these programmes in areas related to Urban Studies. There are master's programmes in

architecture that sometimes provide modules related to urban design but not as part of the core modules.

2 "Marketing manager" is a professional position specialising in understanding how to make people buy certain products. Most marketing managers hold an MBA. They use advertising mainly to ensure that companies allocate their products in the market successfully. A marketing manager must have the capacity to transform complex processes and objectives into simple and sellable products.

6 Towards a theory of urban design under neoliberalism design

Introduction

This book began by stating that neoliberalism has wholly subsumed urban design and that it has neglected its original ethos of designing cities for living by aligning its theory and practice with the objectives of neoliberalism. To test this virtual object, the book has focussed on how urban designers have become instrumental actors in the commoditisation of urban development, thereby undermining their responsibility to develop just and inclusive cities. In this final chapter, I reflect on how the evidence presented in this book serves to construct a preliminary set of theoretical reflections on the features and contradictions of urban design under neoliberalism. Although I have focussed on Santiago de Chile, I believe that many of these reflections apply to other neoliberal metropoles around the globe, especially those in the southern hemisphere. Together, these reflections contribute to the development of a strategy for revolutionising urban design by stripping neoliberal ideologies from the practices of urban design and by elaborating on an autonomous theory for informing and defining its practice, thus rethinking city making in a way that promotes new paradigms, principles, and a new ethos.

Of course, this book could not hazard to define a precise agenda, so I have supported those who are theorising to define tactics and strategies to eradicate neoliberalism. Urban designers need to make use of a revolutionary strategy to liberate themselves as potential actors in the creation of better cities for all. Under neoliberalism, the ethics, practices, and theory of urban design in Santiago have been transformed into a profit-oriented disciplinary field that commodifies urban processes for the sake of capital gains and to increase the exchange value of urban products. To the best of my understanding, the neoliberalisation of urban design is an aberration, and in an attempt to combat this, the present chapter outlines a set of options that could lead towards the revolution that is needed to overcome urban design under neoliberalism.

A definition: urban design under neoliberalism

Throughout this book, I have made visible the relationship between urban design and neoliberalism, using Santiago as an example. The implementation of neoliberalism in Chile involved a series of political-economic practices that were intended to ensure that the country's social relations would be profitable. Under Augusto Pinochet's dictatorship, as instructed by Milton Friedman, the transformation of Chile into a free-market champion followed a non-gradualist pattern of monetary reforms. This transformation reduced the state's authority in public affairs, led to the privatisation of social services, and opened Chile's borders to international capital. Consequently, in 1979, urban development adopted neoliberal principles, which resulted in the transformation of urban land and urban space into assets to be exploited for capital accumulation and wealth creation.

The importance given to property rights in order to ensure land's availability for trading cannot be understated. Before 1979, access to land was viewed as a right, and issues pertaining to the occupation of private property were resolved by the state. After the implementation of neoliberal policies and the PNDU of 1979, however, the state began defending property-owners, hoping to drastically increase the number property-owners in Chile. This was one of the justifications for the urban sprawl policy, which altered urban development in Santiago by increasing the availability of urban land to ensure that everyone had the opportunity to own a piece of it. Thus, urban design began adopting free-market reasoning by giving priority to increasing the exchange value of land and following the supply-and-demand rule.

By declaring land to be a non-scarce resource, extending the urban limits of Santiago, and defining urban development according to market trends, neoliberalism in Chile succeeded 42 years after its introduction. It transformed the social contract and established profit as an engine of progress. Recently, several scandals have illustrated how neoliberalism coerced several Chilean democratic representatives through a scheme by which companies provided funds to politicians and politicians, in turn, facilitated the promulgation of laws and ad hoc regulations designed to strengthen entrepreneurial freedom and private property defences. "This has been the case for land regulations (Caso Caval), fishing (Ley de Pesca), international conflicts (Piñera and Bancard), pension schemes (Grupo Penta), education (Reforma educacional) and environment (Minera Dominga) just to mention a few recent cases" (Vergara-Perucuch in Boano & Vergara-Perucich, 2017, p. 21).

In theory, neoliberalism is a concrete utopia based on the unlimited exploitation of the working class by a dominant elite (Bourdieu, 1998). This concept has dominated urbanisation revenues in Chile since the early years of *urbanismo*. As Janoschka and Hidalgo (2014) observed, the trans-formation of citizens into consumers has created a profit-oriented society, the consequence of which is a neoliberal city. Consequently, urban design adopted new demands and started to design cities in ways that fulfilled the requirements of customers and their purchasing power rather than attending to the rights of citizens.

To briefly define urban design under neoliberalism, it is a mode of spatial production whose goal is to extract profit from urban products. This profit is realised when urban products are priced much higher than the capital invested in them, in which case the value of urban products is not related to its fundamentals (e.g., land price, labour wages, and construction material costs). Various methods of urban design under neoliberalism are intended to reduce the time taken to design good cities, which reduces the quality of urban projects. Part of this problem is the standardisation of design, which maximises the value extracted from the production of space, thus ensuring that every spatial decision increases the rent paid to those who invested in the project. I would like to point out that profit in the case of urban design under neoliberalism refers to the profit of capitalists as well as that of politicians. Urban design under neoliberalism has also been useful in helping the political class to produce urban solutions to social urgencies in collaboration with the capitalist elite when doing so is beneficial in terms of obtaining votes. Consequently, the public-private partnership has become standard practice because public and private elites alike are closely related in terms of their interests.

Under neoliberalism, economic growth, employment, and production have been the primary goals of governments and enterprises. So, it is expected that certain activities, such as urban development projects, and their eco-nomic profitability are subject to exploitation for the sake of a hegemonic class. In practice, urban design under neoliberalism follows a rigid set of economic rules, which are arranged into three key principles:

a)   *Localisation defines the quality of developments*: In housing production, real estate marketers have precise definitions of typologies for buildings based on the location of the plot, which defines the level of income of their potential customers. Thus, based on these two precepts, the design of these buildings must fulfil specific criteria in terms of price, building materials, colour, shape, number of parking spaces per flat, and physi-cal dimensions. In social housing, the criteria are more straightforward:

everything must as cheap as is permitted by building regulations. In this case, any improvement to social housing must be funded by the government through subsidies.

b) *Internal rate of return:* Another guideline is the IRR, which defines how much the investor (either public or private) can spend on innovations in urban design without ensuring high rates of return. Thus, if an investor agrees to allow an urban designer to implement an innovative design in a building project, it must be expected that these innovations will not result in earnings lower than those expected to be acquired via a previously used design.

c) *From the private to the public space:* An accepted definition of urban design in Santiago in the literature does not exist because the city is generally planned from the private to the public sector. While there are some exceptions to this pattern (e.g., infrastructure projects), it is much more common to see architects develop public spaces only as extensions of private projects. In Santiago, there are no urban design projects; there are only urban projects, which are developed under architecture criteria, just on a larger scale. Urban design under neoliberalism depicts a collection of architectural objects that populate the urban tissue.

These conditions strain the imagination and creativity of urban designers. The economic vision of urban investment has neglected the value of urban design, and innovations in urban design are now scarce as a result. This is because the role of the state in urban development is to finance initiatives but not to design them. The situation is worse yet for urban designers working for big urban development or real estate companies whose primary goal is to generate profit. In this case, to be innovative is to take a risk with capital that does not belong to the practitioner. In this way, urban design has become a matter of business instead of a method for organising cites to benefit their inhabitants.

Urban design under neoliberalism oversimplifies the complexities of the city as a way of controlling its development such that the revenue of spatial transformations is maximised, thus ensuring the reproduction of capital. Thus, the goal of urban design under neoliberalism is not to create a good city but to maximise the capital accumulation of a city. Because of this, urban design is void of creativity, having been de-theorised, redefined by marketing and profitability, and abstracted to frequencies and statistics that allow for projects to be designed based on nothing but tables and charts. The use value of urban spaces is considered only when it increases the exchange value of an urban product. Paraphrasing Marx, the control over spatial production is another method of domination that represents the power of a neoliberalist class. Urban design under neoliberalism, thus, is an instrument of social domination.

The fierce defence of private property rights eases the process of capital accumulation through urban development. In this regard, the financialisation of housing, which is one of the consequences of neoliberalisation, has created a complex scenario for shaping a city. Therefore, the purpose of innovation in urban design under neoliberalism is not to create better urban design methods but to find new ways to make more money from urban development processes. It would not be surprising if, in the near future, urban designers begin to specialise in financial innovation, finding design methods that aim solely to transform urban space into a long-term fixed asset with a financial efficiency similar to that of stock and bond markets. This future represents the final negation of the artistic features of urban design. If the pursuit of beautiful cities is endangered under neoliberalism, the financialisation of urban design would cause its extinction.

Financialisation emerges when urban processes are substantially influenced by monetarism. The adoption of monetarism as a primary theoretical source for urban design under neoliberalism has caused an internal contradiction: urban design is restricted to developing lucrative spaces, but in order to achieve this end efficiently, urban design relies on methods borrowed from economics while renouncing the development and deepening of its own methods. However, a contradiction emerges when creativity and imagination are limited by the interests of capitalists and their expectations for profit. Hence, it may be said that urban design under neoliberalism is a spatial specialisation of the monetarist theory of economic disciplines. As an economic instrument, urban design under neoliberalism is an efficient way to cluster groups of consumers into spatial units based on their socio-economic status in order to improve targeting strategies and to more efficiently allocate urban products.

As the case study in this book suggests, urban design under neoliberalism tends to generate monotonous cities based on the reproduction of the same typologies of buildings and streets which are considered successful only because of their profitability, thus reducing the importance of design in the processes of spatial production. The evidence suggests that urban design under neoliberalism strategically reproduces standardised aesthetic proposals that ensure the efficiency of capital investments. As part of its commitment to ensuring profitability, urban design under neoliberalism has developed an aesthetic based on certain principles of the modern movement that have been adapted to increase the profitability of building projects.

Paraphrasing Mies Van der Rohe, regarding urban design under neoliberalism, *less is more*. That is, lower costs equal greater earnings. This is a principle that follows the rules of the internal rate of return, which directly influences the typology of a space. Indeed, this criterion is based on an adaptation of the ideas of supply and demand and the break-even analysis for determining the point at which an investment's revenues outweigh its costs.

The break-even analysis in urban design promotes the employment of economies of scale for urban design, which means that a successful urban product that has already been designed is simply replicated throughout the city (with slight differences). This strategy reduces the time required for designing, thereby increasing revenue. Urban design under neoliberalism borrows from economic theories, such as the previously mentioned IRR and NPV, to inform its spatial strategies to decide what can be constructed and what cannot, the economies of scale, the monetarist theory, the cost-benefit approach, and the generalised acceptance of using a positivist framework to interpret urban problems.

Cities under neoliberalist rule, which now have more wealth, technology, and data than ever before, have been transformed into generic architectural objects with repetitious typologies which are efficient as commodities which are effective at generating revenue through their commercialisation but weak in their aesthetic proposals and innovative qualities. Therefore, urban design under neoliberalism is an extremely conservative way of developing a city. Although the number of housing units and the availability of amenities in Santiago have increased, the magnitude of new urban problems makes them difficult to resolve without implementing a radical approach that redefines urban design.

Despite the macroeconomic success of Chile, the data reported in this book reveal that after 40 years of neoliberalism, income inequality has not improved, which may be reflected in the levels of urban segregation in Santiago. Many will argue that neoliberalism drastically reduced poverty and increased the per capita GDP in Chile. Indeed, this has happened over the past few decades, but the same is true for non-neoliberalised countries. In fact, most advanced societies in the world – including Norway, Denmark, Sweden, Canada, Finland, Iceland, and a long list of other nations – reached a state of development while avoiding the dogmatic ways adopted by Chilean neoliberalism. These countries instead followed a system in which the role of the state and civil society in shaping cities was not controlled by free-market economics. Chilean citizens are now realising that neoliberal ideals are mere illusions, that the neoliberalised social (in)security system is defective, and that housing will soon become a central issue in the political agenda in which private property and market-driven solutions will be identified as structural problems that need to be resolved.

The ethical conflict of urban designers under neoliberalism

Under the neoliberal regime, the disciplinary field of urban design has re-aligned its ethos in accordance with economic theories through a series of profit-oriented modes of production to ensure that urban development is financially lucrative. Urban design under neoliberalism has a deep complexity that this research has tried to outline and explore in order to better understand its logic and reveal its contradictions. Urban design under neoliberalism triggers ethical paradoxes for developing urban projects. When referring to ethics, I refer to the discernment between right and wrong regarding the decisions made by people related to a specific activity (Frey & Wellman, 2003). In the case of urban design in Chile, ethics are based mainly on the production of space and its outcome: the city (Munizaga, 2014).

Conversely, in Marxist approaches, ethics depict individuals' awareness of and responsibility for improving society (Lukács, 2014). However, for Marx, ethics was also a moral illusion governing social practices in which a hegemonic class defines what is right and wrong. Indeed, the late works of Marx questioned the moral stance of ethics as an illusion and an ideological construction (Kain, 1988). By embracing a critical Marxist approach to ethics, urban design refers to the decisions taken when shaping cities in order to develop spaces that benefit the majority, thus producing spatial equality and allowing inhabitants to embrace their built environment with a sense of belonging.

To summarise this discussion, there are two ethical streams of urban design: urban design for the good of the city as a complex social construction and urban design under neoliberalism. The first stream is defined by an idealisation of what urban designers' goals should be when shaping cities. In this sense, any ethical reflection on urban design under neoliberalism needs to make clear the difference between a profitable city and a good city. Under neoliberalism, it is expected that a good city will be profitable, growing, and dynamic. From a neoliberal perspective, a profitable city *is* a good city.

Henri Lefebvre (2003) argues that ethical illusions in urban practices, such as urban design, may lead to a revolutionary practice. Theories of aesthetics and ethics can be used together to create strategies for removing capitalism from social practices (Kain, 1988). Furthermore, critiques of the ethos of neoliberalists may foster the development of revolutionary tactics against urban design under neoliberalism. In fact, such tactics have already been attempted by UKAMAU, MilM2, AriztíaLAB, Movimiento de Pobladores en Lucha, Movimiento por una Vivienda Digna, and Grupo TOMA, to mention just a few examples. Eamonn Canniffe argues that urban disciplines may take ethical positions regarding the exploitation of people and land by capitalist modes of production. These positions do not need to be expressed in public value judgements but may be inherent in the very practice of designing spaces and outlining the process of spatial production (Canniffe, 2006b). Furthermore, Karsten Harries (1997) argues that the ethical function of design disciplines is to move towards the ideal of a better life; to preserve the utopia as a seed of utopian longings; and to create an expectation in society for a different, better world, which is very much needed in the current state of our civilisation.

The findings presented here suggest that the ethical transformation of urban design in Chile started in 1979 when the national policy of urban design determined that the market would determine the best way to use land. In other words, the use of land hinged on the interests of investors with no regard for any long-term plans for the city. Indeed, the long-term plan for Santiago was not considered.

This policy changed the way the processes of the production of space were interpreted by heavily stressing the extraction of surplus value from urban development. As a consequence, urban design was no longer considered an important discipline. Public space was considered a result, and, given that it did not yield profits, its design was an afterthought of urban developers. The neoliberal revolution dismantled the strong urban planning apparatus of the Chilean state, which was composed of a series of public institutions tasked with defining the urban form and organising urban development projects. In order to survive the neoliberal revolution, urban designers were forced to change their ethos and embrace financial markets as their main inspiration.

Figure 6.1 organises the ethical contradictions that emerged from studying urban design under neoliberalism. The limitations imposed on the practice of urban design include the fragmentation of institutions which had previously coordinated urban processes, the scarcity of resources available for developing public spaces in relation to private spaces, the planning of objectives for urban space according to short-term schedules, the excessive trust in the free market's invisible hand for regulating urban development, and the rule of private property mentioned previously. Regarding the ethos of urban planning and the conflicts surrounding it, there is resistance to innovation growing out of the constraining framework of profitability caused by a lack of discussions about the aesthetic image of the city combined with the dominance of quantitative methods for informing decisions while rejecting complex sources of data, such as participative observations, ethnography, grounded theories, and qualitative methods.

Furthermore, urban designers in Santiago naturalised the public-private partnership when developing cities. Urban design under neoliberalism idealised an efficient way of developing the city in which the state provided funds and private companies executed the work in a way that ensured the development process would yield earnings. Within this scheme, the state develops projects that are relevant to the country, and private companies participate in the spatial development of the country because the state guarantees profitability. The problem with this is that the state has no control over the allocation of profits.

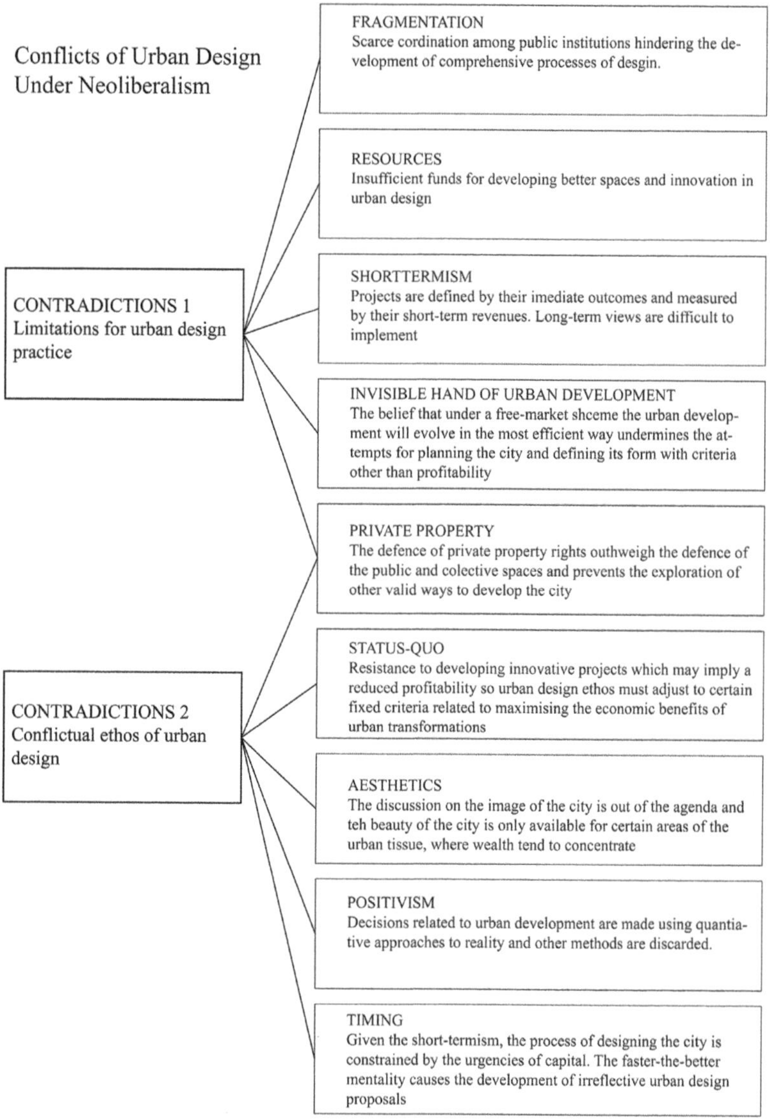

Conflicts of Urban Design Under Neoliberalism

FRAGMENTATION
Scarce cordination among public institutions hindering the development of comprehensive processes of desgin.

RESOURCES
Insufficient funds for developing better spaces and innovation in urban design

CONTRADICTIONS 1
Limitations for urban design practice

SHORTTERMISM
Projects are defined by their imediate outcomes and measured by their short-term revenues. Long-term views are difficult to implement

INVISIBLE HAND OF URBAN DEVELOPMENT
The belief that under a free-market shceme the urban development will evolve in the most efficient way undermines the attempts for planning the city and defining its form with criteria other than profitability

PRIVATE PROPERTY
The defence of private property rights outhweigh the defence of the public and colective spaces and prevents the exploration of other valid ways to develop the city

STATUS-QUO
Resistance to developing innovative projects which may imply a reduced profitability so urban design ethos must adjust to certain fixed criteria related to maximising the economic benefits of urban transformations

CONTRADICTIONS 2
Conflictual ethos of urban design

AESTHETICS
The discussion on the image of the city is out of the agenda and teh beauty of the city is only available for certain areas of the urban tissue, where wealth tend to concentrate

POSITIVISM
Decisions related to urban development are made using quantiative approaches to reality and other methods are discarded.

TIMING
Given the short-termism, the process of designing the city is constrained by the urgencies of capital. The faster-the-better mentality causes the development of irreflective urban design proposals

Figure 6.1 Summary of findings that reflect the conflicts of urban designers under neoliberalism.

Source: Author.

A highly profitable city is not necessarily better for everyone, though it surely is best for the few who can afford to invest in urban projects. The segregation, fragmentation, and gentrification of society in Santiago mean that a city that is good for everyone is an illusion. As illustrated in this thesis, Santiago is a divided city with deep contrasts. One city – a wealthy, green, vibrant, luxurious city – rests at the top, and the other city – an impoverished, dry, cheap city – lies at the bottom. It is likely that Santiago's urban designers would prefer a more just city in which everyone has access to public goods and high-quality living spaces. However, this is only an aspiration. Urban design under neoliberalism has created a compelling illusion of the good city. Thus, the question for urban designers remains unsolved as to how they can implement urban design under neoliberalism while relieving the public of social injustices and reconcile free-market dogmas in pursuit of the truly good city in the Latin American context.

Urban designers are aware of the neoliberalisation that dominates their discipline and the effect it has on urban development. As the evidence provided in this book has suggested, urban designers know very well that their practice has been distorted to become profit-oriented; still, they continue using the same methodological apparatuses to shape city spaces and urban life because the decision-making power remains in the hands of an oligarchy. This implies that the first step towards emancipating urban design from neoliberalism is already being taken. The consciousness of urban design practitioners regarding the crisis that their disciplinary field is experiencing because of profit-driven logic is an imperative movement that has started with the recognition that the neoliberal condition jeopardises their ethical integrity.

A reflection has emerged from the critical context of violence, dictatorship, and human rights violations in which neoliberalism was implemented. It is vital to note that urban design under neoliberalism emerged from a hostile social context. In order to survive the process of neoliberalisation, urban designers changed their ethos. Their survival was in question in relation to not only their lives but also their livelihood. Basically, urban designers depend on spatial transformation to make a living, which is yet another way in which control has been imposed on them. In order to be hired for designing projects, urban designers must fulfil the requests and expectations of investors. This condition exists because urban design under neoliberalism was created not by urban designers but by investors, capitalists, speculators, the dictatorial state (and the traumatised pragmatic post-dictatorial governments), and the economic elite. The fragmentation of urban processes and entities is critical, but so is the fragmentation of urban designers, whose disciplines have been assaulted, reshaped, and profoundly transformed by capitalists. Indeed, the most convenient scenario for capitalists is fragmented and disarticulated relationships among urban designers. An alienated disciplinary body is ideal for capital in every aspect of the economy. I believe that if urban designers do not organise to promote their demand to remove profit-driven goals from their practice, capitalists will maintain their hegemonic position and neoliberalism will continue to be the leading influence on urban design practices.

Recalling the topic of the ethical contradictions in urban design under neoliberalism, although there is a complicity by omission (urban designers recognise the problems of neoliberalisation but have not organised a massive resistance against it), an awareness of the problems and contradictions of urban design under neoliberalism seems to be present in the critical discourse of urban design practices in Santiago. The collective organisation of urban design practitioners to share their views is a fundamental step to be taken in redefining the ethics of urban design in Santiago and clearing the neoliberal fog that has blurred their ethos.

Assessing urban design under neoliberalism

In this field, the actions of building companies and their coordination with the government add dynamics to the present cycle. The democratic government is interested in this cycle because it can extract a significant amount of financial capital through taxes on the surplus value created by the building companies.

In the diagram in Figure 6.2, the praxis and ethos of urban development are on the right side. The coordinate (Urban; Praxis) focuses on the everydayness of urban design under neoliberalism as a daily experience. The market for spaces and urban products is sold to generate urban experiences. In this coordinate, the idea of the good city is strongly influenced by the goals of the market. Thus, the urban experience and the profitability of spatial production are linked. This link is derived from the fact that, on the left (Market; Praxis), supply is organised, while demand comes from the right (Urban; Praxis). Through the urban experience, processes involving the consumption of urban products are developed as fabricated by urban design under neoliberalism. Finally, the coordinate (Urban; Ethos) contains the urban designers, who are directly linked to the demands of the government and whose livelihoods depend on the jobs created by building companies and investors. Urban designers have to adopt the goals of the government as their own. Thus, the spaces within good cities are infused with the interests of building companies and the goals of the government. In urban design under neoliberalism, the idea of the good city is based on profitable outcomes and interventions that create surplus value by activating cycles in the consumption of urban products in everyday life.

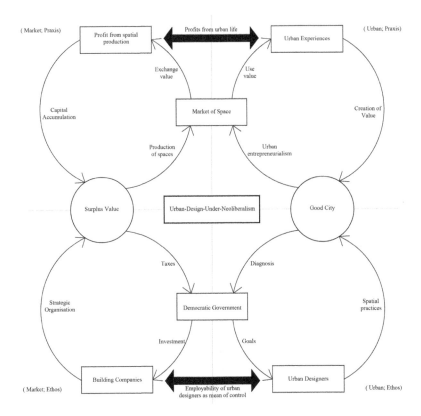

Figure 6.2 Diagram of the composition of urban design under neoliberalism.

Source: Author (inspired by David Harvey (1978)).

The ethics, theories, and practices of urban design are the three main elements that this diagram attempts to organise. By organising these elements in this way, I suggest that urban designers are subsumed by the goals of the government and of profit-oriented institutions. The ethics of urban design under neoliberalism are contradictory from my ontological position, as I have embraced an understanding of the goals of urban design as the pursuit of better spaces for all. A change in the practice of urban design in a neoliberal regime will require a revolutionary strategy for liberating its ethos and redefining it with a different framework. I believe that profitability is not bad per se, but the entire transformation of a disciplinary field by an exogenous theory is destructive, especially when it aims to take advantage of the skills and capacities of the practitioners working within that field.

I was trained as an urban designer in Santiago. During my years as a student, the modules I studied fostered an imaginative and creative attitude towards designing cities. If a project was too simple or ignored social contributions, it was given a failing grade. The worst grades were given to projects designed similarly to the way the buildings of Santiago are designed. All those years of critical formation disappeared when I started to work in the real world. My imagination was constrained, and my criticisms were ignored.

After school, urban designers enter a productive phase in which the years of critical formation spent at university are disregarded. If urban design requires a revolution for its emancipation, the manifesto for that revolution is already written in the imaginative and thought-provoking realm of architecture schools, where urban designers are trained in Santiago and in other cities throughout the world. The challenge is bringing that evocative and critical way of reading and thinking from universities to the processes of producing spaces in cities. A new virtual object is required in the push for a revolutionary agenda of urban design.

It can be said that urban society in Santiago has been neoliberalised by the collaboration of urban design under neoliberalism, a complicit disciplinary field that has been tailored to suit capital-based objectives. The ethical crises of urban design in Santiago are represented by supporting the pragmatist praxis, abandoning utopian thinking, and reducing imaginative capacity regarding what the market allows. Because of these crises, urban design is trapped in a blind field constructed by neoliberal ideology. Urban designers in Santiago dislike the unjust and uncreative city that has been developed under neoliberalism – or at least they believe it is not a successful model from the point of view of the majority of its inhabitants.

This makes me think that there is fertile soil in Santiago for cultivating a discussion about breaking the connection between urban design and neoliberalism. I believe that a disciplinary revolution is needed immediately. However, a possible revolution of urban design under neoliberalism needs a critical practice informed by a radical theory that is capable of transforming the ethics of urban designers to generate a self-sufficient discipline for developing good cities. Revolutionising urban design under neoliberalism must be a project of disciplinary autonomy.

A virtual object for further research: the solidary urban design practice

To contest urban design under neoliberalism, I propose that a counter-virtual object capable of offering theoretical contestation should be developed. I call this counter-virtual object "the solidary urban design practice". The solidary urban design practice serves as a virtual object that critiques the neoliberalisation of Santiago while offering an alternative way of understanding of urban life. The concrete utopia of a post-neoliberal city is absent from the discussion. From my perspective as an urban designer trained in Santiago, the definition of a solidary city starts with a struggle to make everyday life in cities better for all. While urban design is meant to provide spaces for society that benefit the majority, the outcomes of urban design under neoliberalism are individualistic and alienating and facilitate the spatial division of classes. The solidary urban design practice would offer a theoretical object that discusses the development of elaborate spaces to achieve outcomes that are opposite to the values and principles of neoliberalism.

The solidary urban design practice is a virtual object which does not yet exist but which needs to be developed in Santiago. The common good, rather than capital, would be at the centre of the decision-making process in urban design projects in this virtual elaboration of the solidary urban design practice. This implies that the city must be designed in a way that reduces asymmetric access to high-quality urban spaces. In practice, this implies the dual challenge of increasing the quality of public spaces across the city while ensuring that all inhabitants can commute smoothly from one side of the metropolis to the other.

In the solidary city (the outcome of this new practice), affordability and access to urban life – namely housing, transport, leisure, and services – are basic principles for starting, but they are not enough. Today, life in cities like Santiago is very expensive. Affordability means that people can pay for something without straining themselves financially. Taking this idea further, the solidary city aims to produce what Neil Brenner, Margit Mayer, and Peter Marcuse have referred to as cities for people, not for profit. No citizen can be excluded from urban life, and no part of urban design can be subjugated by the interests of capitalists. This kind of spatial justice is one of the core values of a solidary city. In order to ensure that spatial justice prevails in a city, the solidary urban design practice must create institutions and agencies capable of organising people for discussing and defining the urban form. Therefore, the solidary city results from an imagined future of urban life in which the whole population is compelled to participate in its design, thus making its spatial conception collective. These ideas can be extracted from the evidence collected during this research.

In an effort to launch this debate, I have developed six reflections, offered as potential strategies that may help to subvert urban design under neoliberalism, break its the connection between urban design and neoliberalism, and outline the solidary urban design theory:

1   Urban designers constitute a body of specialists whose central ethical commitment is to shape cities for good, or, as I have suggested earlier, to

practice according to the theory of solidary urban design. Urban design must be re-politicised. In order for this to happen, urban designers need to shift their focus from the elite to ordinary people by embedding their practices in the spatial scarcities of the many people who live in precarious conditions. Those affected by urban design under neoliberalism are the primary target of a subversive strategy to overthrow this mode of developing spaces. Acting as social catalysts, urban designers must use their knowledge for the service of the civil society that endures segregated, gentrified, fragmented, alienated, indebted, and isolated lives.

2   Urban designers should evaluate and recognise the importance of producing a theory of their practice and building an epistemology of the art of shaping cities. While this discipline remains subjugated to econometric, sociological, and political theories, its own discipline-specific theoretical reflection could provide ideas for planning its emancipation from neoliberalism. Theoretical autonomy is crucial because it emerges from the ethics and expectations of practitioners while avoiding the influence of external disciplines that may see urban design as something to be manipulated and exploited for monetary gains.

3   Urban designers should resume the use of utopian ideals for building imaginative and desirable futures for an alternative world. Urban designers should return to their creative nature by proposing alternative futures for society. This was one of the messages Henri Lefebvre delivered throughout his works. The concrete utopia provides an example of a feasible future spatial scenario. The main goal of this utopian way of thinking is to develop prospects that society will want to live in, thus convincing people to demand spatial changes in their everyday lives. Furthermore, the concrete utopia may become a political tool for promoting urban design and using it to offer people a better future, one in which neoliberalism is no longer needed.

4   Urban designers must enter the political arena as mayors, representatives, social leaders, and other positions of leadership. This is fundamental to a revolution because the struggle to overthrow urban design under neoliberalism requires the significant transformation of laws, regulations, bureaucratic processes, and market rules.

5   The previous point allows for the proposal of a structural modification – specifically, private property must be adapted according to the previous social role of the space. Private interest cannot be given priority over the common good. Specific spaces in cities must be assigned for inverting segregation, gentrification, fragmentation, and financialisation. One of the primary goals of the solidary urban design theory is to consolidate a methodology in which the collective is more important than the individual.

6   The solidary urban design practice must take a revolutionary approach to the discipline of urban design with the aim of contesting and eradicating urban design under neoliberalism. This revolution has already been mentioned in this chapter and refers to a profound change in everyday life in which the entire city is reorganised for the sake of the community; urban design plays a key role in providing spaces for this goal to be worked toward. It is a radical agenda for change that requires the involvement of several disciplines, such as urban design, architecture, politics, economics, sociology, geography, etc.

The solidary urban design practice adopts the principles of the right to the city, updated according to the problems caused by neoliberalism. In other words, the implementation of the right to the city is one of the main objectives of the solidary urban design. The exploration of the solidary city may serve to organise widespread alternative practices in Santiago around the concept of the concrete utopia, reflecting on a future for the city that is desirable, possible, and necessary. The virtual object of the solidary city offers the possibility of developing further research to answer several key questions: What does a solidary city mean for urban designers and the general public in Santiago? What can one expect to find in a solidary city? What does it look like? What does everyday urban life look like in a solidary city? The construction of this virtual object may serve as a strategy for resituating the importance of the utopia and of the imagination in reflecting the futures of cities.

Bibliography

Addie, J. P. D. (2008). The rhetoric and reality of urban policy in the neoliberal city: Implications for social struggle in over-the-rhine, Cincinnati. *Environment and Planning A, 40*(11), 2674–2692. https://doi.org/10.1068/a4045

Agostini, C. A., Hojman, D., Román, A., & Valenzuela, L. (2016). Segregación residencial de ingresos en el Gran Santiago, 1992–2002: Una estimación robusta. *Eure, 42*(127), 159–184. https://doi.org/10.4067/S0250-71612016000300007

Aguilar, R., Oliva, C., & Laclabere, S. (2016). IMAGEN URBANA Y PODER. INFLUENCIAS DEL NEOLIBERALISMO Y LA GLOBALIZACIÓN EN LA REPRESENTACIÓN ARQUITECTÓNICA DE SANTIAGO DE CHILE. *Revista Diseño Urbano Y Paisaje, 31*(1), 24–29.

Allen, J., & Pryke, M. (1994). The production of service space. *Environment and Planning D: Society and Space, 12*(4), 453–475. https://doi.org/10.1068/d120453

Amin, A. (2002). The economic base of contemporary cities. *A Companion to the City,* 115–129. Retrieved from www.amazon.com/dp/0631235787

Amin, A. (2006). The good city. *Urban Studies, 43*(5–6), 1009–1023. https://doi.org/10.1080/00420980600676717

Amin, A., & Thrift, N. (2002). Cities: Reimagining the urban. *Urban Policy and Research, 21*(2), 217–222. https://doi.org/10.1098/rsnr.2002.0171

Amunátegui Solar, D. (1938). Don Andrés Bello Enseña a los Chilenos a Narrar la Historia Nacional. In *Anales de la Universidad de Chile* (pp. 33–34). Santiago: Universidad de Chile.

Arancibia Clavel, P., & Balart, F. (2007). *Sergio de Castro, el Arquitecto del Modelo Económico Chileno.* Santiago: Biblioteca Americana.

Assadi, F., Pulido, F., & Zapata, D. (2008). Edificio gen. *ARQ, 1*(96), 48–53.

Atria, F. (2017). [VIDEO] *Fernando Atria: "El Transantiago es un monumento al neoliberalismo chileno" – El Mostrador.* Retrieved November 16, 2017, from www.elmostrador.cl/noticias/multimedia/2017/02/10/video-fernando-atria-el-transantiago-es-un-monumento-al-neoliberalismo-chileno/

Atria, F., Larrain, G., Benavente, J. M., Couso, J., & Joignant, A. (2013). *El otro modelo.* Santiago: Random House Mondadori.

Bakunin, M. (1975). *El principio del estado. Universidad Complutense de Madrid.* Retrieved from www.ucm.es/info/bas/utopia/html/elprinci.htm%5Cnhttp://sagunto.cnt.es/wp-content/uploads/2010/12/El-principio-del-estado.pdf

Banerjee, T., & Louaitou-Sideris, A. (2011). *Companion to urban design*. London and New York: Routledge. https://doi.org/10.4324/9780203844434

Barnett, J. (1982). *An introduction to urban design*. New York: Ledgebrook.

Barnett, J. (2011). A short guide to the 60 newest urbanisms. *Planning, 77*(4), 19–21.

Benévolo, L. (1985). *La ciudad en la historia*. Barcelona: Paidos Iberica.

Bengoa, J. (1996). *Historia del pueblo Mapuche*. Santiago: LOM Ediciones.

Bengoa, J. (1999). *Haciendas y Campesinos*. Santiago: SUR Ediciones.

Bentley, I., & Butina, G. (1976). What is Urban Design? In *Towards a Definition*. Oxford: Urban Design Forum.

Biddulph, M. (2012). The problem with thinking about or for urban design. *Journal of Urban Design, 17*(1), 1–20. https://doi.org/10.1080/13574809.2011.646251

Boano, C. (2017). *Ethics of a potential urbanism: Critical encounters and the built environment*. London: Routledge.

Boano, C., Lamarca, M. G., & Hunter, W. (2011). The frontlines of contested urbanism mega-projects and mega-resistances in Dharavi. *Journal of Developing Societies, 27*(3–4), 295–326. https://doi.org/10.1177/0169796X1102700404

Boano, C., & Talocci, G. (2014). Fences and profanations: Questioning the sacredness of urban design. *Journal of Urban Design, 19*(5), 700–721. https://doi.org/10.1080/13574809.2014.943701

Boano, C., & Vergara-Perucich, F. (2017). *Neoliberalism and urban development in Latin America*. London and New York: Routledge.

Boano, C., & Vergara-Perucich, J. F. (2016). Bajo escasez. ¿Media casa basta? Reflexiones sobre el Pritzker de Alejandro Aravena. *Revista de Arquitectura, 21*(31), 37–46. https://doi.org/10.5354/0719-5427.2016.42516

Boas, T. C., & Gans-Morse, J. (2009). Neoliberalism: From new liberal philosophy to anti-liberal slogan. *Studies in Comparative International Development, 44*(2), 137–161. https://doi.org/10.1007/s12116-009-9040-5

Bourdieu, P. (1998, December). Utopia of endless exploitation: The essence of neoliberalism. *Le Monde Diplomatique*, 1–6.

Bowen, G. A. (2009). Document analysis as a qualitative research method. *Qualitative Research Journal, 9*(2), 27–40. https://doi.org/http://dx.doi.org/10.3316/QRJ0902027

Boza, C., Castedo, L., & Duval, H. (1983). *Santiago, estilos y ornamento: Estilos y ornamento*. Montt Palumbo.

Braun, J., Braun, M., Briones, I., & Díaz, J. (2000). Economía chilena 1810–1995, Estadisticas historicas. *Documento de Trabajo, 187*, 369. Retrieved from www.economia.puc.cl/docs/dt_187.pdf

Brenner, N. (2013). Theses on Urbanization. *Public Culture, 25*(1), 85–114. https://doi.org/10.1215/08992363-1890477

Brenner, N. (2014a). *Implosions/explosions: Towards a study of planetary urbanization*. Neil J Brenner: 9783868593174: Amazon.com: Books. Retrieved from www.amazon.com/Implosions-Explosions-Towards-Planetary-Urbanization/dp/3868593179/ref=sr_1_1?ie=UTF8&qid=1434621477&sr=8-1&keywords=implosions+explosions

Brenner, N. (2014b). Introduction: Urban theory without an outside. *Implosions/Explosions: Towards a Study of Planetary Urbanization*, 15–31.

Brenner, N. (2016). The hinterland urbanised? *Architectural Design, 86*(4), 118–127. https://doi.org/10.1002/ad.2077

Brenner, N., & Elden, S. (2001). Henri Lefebvre in contexts: An introduction. *Antipode, 33*(5), 763–768. https://doi.org/10.1111/1467-8330.00215

Brenner, N., & Theodore, N. (2002). Spaces of neoliberalism: urban restructuring in Western Europe and North America. *Antipode.* https://doi.org/10.1002/9781444397499. ch4

Brenner, N., Madden, D. J., & Wachsmuth, D. (2011). Assemblage urbanism and the challenges of critical urban theory. *City, 15*(2). https://doi.org/10.1080/1360 4813.2011.568717

Brenner, N., Marcuse, P., & Mayer, M. (2012). *Cities for people, not ofr profit.* London and New York: Routledge.

Brenner, N., Peck, J., & Theodore, N. (2012). *Alternatives of neoliberalism.* London: Bedford Press.

Brunner, K. (1939). *Manual de Urbanismo.* Bogota: Ediciones del Consejo Editorial de Bogota.

Brunner, K., & Meltzer, A. H. (1972). Friedman's monetary theory. *Journal of Political Economy, 80*(5), 837–51.

Bunge, M. (1974). *Treatise on basic philosophy, Vol. I Semantics I: Sense and reference.* Dordrecth-Boston: Reidel Publishing Co.

Burgel, G., Burgel, G., & Dezes, M. G. (1987). An interview with Henri Lefebvre. *Environment and Planning D: Society and Space, 5*(1), 27–38. https://doi.org/10.1068/ d050027

Bustamante, P. (2014). *Chile: Desde la Cosmópolis precolombina del Mapocho a la Cosmópolis actual.* (1 No. 1). WACA: Santiago. Retrieved from https://www. google.cl/url?sa=t&rct=j&q=&esrc=s&source=web&cd=1&cad=rja&uact=8&ve d=0ahUKEwjJiZuIpu7VAhUJjJAKHYoEDhgQFggmMAA&url=http%3A%2F% 2Fwww.waca.cl%2Fpdf%2FPB%2FChile%2520cosmopolis%2520precolombina. pdf&usg=AFQjCNFWRkhZjwX8M17OOawOZGSQClmDdg

Bustamante, P., & Moyano, R. (2014). *EL CUZCO DEL MAPOCHO.* Retrieved August 20, 2015, from http://waca.cl/pdf/PB/ELCUZCODELMAPOCHOResidencia Cancha.pdf

Canniffe, E. (2006a). *Urban ethic: Design in the contemporary city.* London and New York: Routledge.

Canniffe, E. (2006b). *Urban ethic: Design in the contemporary city.* London and New York: Routledge.

Carmona, M. (1998). Design control: Bridging the professional divide, part 1: A new framework. *Ournal of Urban Design, 3*(2), 175–200.

Carmona, M. (2001). *The value of urban design.* Kent: CABE-Commission for Architecture and the Built Environment. Great Britain. Department of the Environment, Transport and the Regions.

Carmona, M. (2009). World class places or decent local spaces for all? *Urban Design International, 14*(4), 189–191.

Carmona, M. (2014a). *Explorations in Urban Design: An Urban Design Research Primer.* London: Ashgate.

Carmona, M. (2014b). The Place-shaping Continuum: A Theory of Urban Design Process. *Journal of Urban Design.* Taylor & Francis. https://doi.org/10.1080/13 574809.2013.854695

Carmona, M. (2015). Re-theorising contemporary public space: A new narrative and a new normative. *Journal of Urbanism, 8*(4), 373–405. https://doi.org/10.1080/17549175.2014.909518

Castells, M. (1997). Citizen movements, information and analysis. *City, 2*(7), 140–155. https://doi.org/10.1080/13604819708900067

Castells, M. (2015). *Networks of outrage and hope: Social movements in the internet age*. New York: John Wiley & Sons.

Cattaneo Pineda, R. A. (2011). Los fondos de inversión inmobiliaria y la producción privada de vivienda en Santiago de Chile: ¿Un nuevo paso hacia la financiarización de la ciudad? *EURE (Santiago), 37*(112), 5–22. https://doi.org/10.4067/S0250-71612011000300001

Celik, Z., & Favro, D. (1988). Methods of urban history. *Journal of Architectural Education, 41*(3), 4–9.

Chadwick, R. (2012). *Encyclopedia of applied ethics*. New York and London: Elsevier Ltd. Retrieved from www.sciencedirect.com/science/referenceworks/9780123739322

Chamboredo, J. C., & Lemaire, M. (1970). Proximité spatiale et distance sociale. Les grands ensembles et leur peuplement. *Revue Française de Sociologie, 11*(1), 3–33.

Chang, H.-J. (2014). *Economics: The user's guide*. London and New York: Penguin Books.

Charnock, G. (2010). Challenging new state spatialities: The open marxism of Henri Lefebvre. *Antipode*, 42(5), 1279–1303.

Chombart de Lauwe, P.-H., Paquot, T., & Tailand, S. (1996). *Un anthropologue dans le siècle*. Paris: Descartes et Cie.

CNDU. (2015). *Política Nacional de Desarrollo Urbano*. Santiago. https://doi.org/http://cndu.gob.cl/wp-content/uploads/2014/10/L4-Politica-Nacional-Urbana.pdf

Cociña, C. (2007). *La Polémica del Costanera Center*. Retrieved May 16, 2016, from www.plataformaurbana.cl/archive/2007/12/20/costanera-center-¿construccion-ilegal/ 2/13

Cociña, C. (2012). *Por qué hemos construido guetos y lo seguimos haciendo | CIPER Chile CIPER Chile" Centro de Investigación e Información Periodística*. Retrieved November 16, 2017, from http://ciperchile.cl/2012/11/14/por-que-hemos-const ruido-guetos-y-lo-seguimos-haciendo/

Cociña, C. (2016). Habitar desigualdades: Politicas urbanas y el despliegue de la vida en Bajos de Mena. *Serie Documentos de Trabajo PNUD-Desigualdad*, (2016/05), 1–21.

COES. (2015). *Encuesta COES. Santiago*: Centro de Estudios para la Cohesion Social.

Coleman, J. S. (1986). Social theory, social research, and a theory of action. *American Journal of Sociology, 91*(6), 1309–1335. https://doi.org/10.1086/228423

Coleman, N. (2015). *Lefebvre for architects*. London and New York: Routledge.

Contardo, J. I. (2014). La recuperación del diseño cívico como reconstrucción de lo local en la ciudad intermedia: El caso de Talca, Chile. *AUS, 2014*(15), 4–8. https://doi.org/10.4206/aus.2014.n15-02

Costes, L. (2010). Le Droit à la ville de Henri Lefebvre: quel héritage politique et scientifique?. *Espaces et sociétés*, (1), 177–191.

Cunliffe, A. L., & Karunanayake, G. (2013). Working within hyphen-spaces in ethnographic research. *Organizational Research Methods, 16*(3), 364–392. https://doi.org/10.1177/1094428113489353

Cuthbert, A. R. (2006a). *The form of cities: Political economy and urban design.* Oxford and Victoria, MA: Blackwell.

Cuthbert, A. R. (2006b). Urban design origins. In A. Cuthbert (Ed.), *The form of cities: Political economy and urban design* (pp. 1–8). London: Routledge.

Cuthbert, A. R. (2007). Urban design: Requiem for an era: Review and critique of the last 50 years. *Urban Design International, 12*(4), 177–223.

Cuthbert, A. (2010). Whose Urban design? *Journal of Urban Design. 15*(3). pp. 443–448. Availabe at: https://doi.org/10.1080/13574809.2010.487816

Cuthbert, A. R. (2011). *Understanding cities: Method in urban design.* London: Routledge.

Daher, A. (1991). Neoliberalismo urbano en Chile. *Estudios Públicos*, 281–299.

Davis, W. (2017, November 16). What is "Neo" about neoliberalism? *New Republic.* Retrieved from https://newrepublic.com/article/143849/neo-neoliberalism

Debord, G. (1994). *The society of spectacle.* New York: Zone Books.

De Mattos, C. A. (1999). Santiago de Chile, globalización y expansión metropolitana: lo que existía sigue existiendo. *Revista EURE-Revista de Estudios Urbano Regionales, 25*(76), 29–56.

De Mattos, C. A. (2002). Santiago de Chile de cara a la Globalización: ¿otra ciudad? *Revista de Sociologia E Política, 19*, 31–54. https://doi.org/http://dx.doi.org/10.5380/rsp.v19i0

De Ramón, A. (2007). Santiago de Chile. Historia. Santiago: Catalonia.

De Ramón, A., & Gross, P. (1985). Santiago de Chile: Caracteristicas histórico ambientales, 1891–1924 (Vol. 1). Santiago: Londres.

De Simone, L. (2012). *COSTANERA CENTER Y LOS MALLESTARES URBANOS: LA URGENCIA DE UN URBANISMO COMERCIAL EN LA PLANIFICACIÓN DE LAS ESTRUCTURAS DE CONSUMO EN LA CIUDAD.* Retrieved May 16, 2016, from http://revistaplaneo.cl/2012/07/02/costanera-center-y-los-mallestares-urbanos-la-urgencia-de-un-urbanismo-comercial-en-la-planificacion-de-las-estruc...

de Valdivia, P. (1545). *Carta de Pedro de Valdivia a Emperador Carlos V.* La Serena, Chile: Archivos Nacionales de Chile.

Díaz, J., Lüders, R., & Wagner, G. (2016). *La República en cifras.* Santiago: Editorial Universitaria. Retrieved from http://cliolab.economia.uc.cl/BD.html

Díaz Osorio, J. (1993). LA RELACIÓN SUELO-HOMBRE EN EL PERÍODO PRECOLOMBINO. *Revista Universum, 8.*

di Girolamo, G. (2015). *Revolucionarios de la vivienda social: La historia de lucha de Ukamau | El Desconcierto.* Retrieved November 16, 2017, from www.eldesconcierto.cl/2015/10/07/revolucionarios-de-la-vivienda-social/

Donoso, F., & Sabatini, F. (1980). Santiago: empresa inmobiliaria compra terrenos. *EURE: Revista Latinoamericana de Estudios Urbanos Y Territoriales, 7*(20), 25–51.

Donoso-Díaz, S., & Arias-Rojas, Ó. (2013). Desplazamiento cotidiano de estudiantes entre comunas de chile: Evidencia y recomendaciones de política para la nueva institucionalidad de la educación pública. *Eure, 39*(116), 39–73. https://doi.org/10.4067/S0250-71612013000100002

Douglas-Irvine, H. (1928). The landholding system of colonial chile. *The Hispanic American Historical Review, 8*(4), 449–495.

Dovey, K., & Pafka, E. (2015). The science of urban design? *Urban Design International, 21*, 1–10. https://doi.org/10.1057/udi.2015.28

Eagleton, T. (2011). *Why Marx was right*. New Haven: Yale University Press.

The Economist. (2017). Santiago's transport system is sputtering: Going nowhere. Retrieved November 16, 2017, from www.economist.com/news/americas/2172 0651-commuters-do-not-want-pay-bad-service-santiagos-transport-system-sputtering

Elden, S. (2001). Politics, philosophy, geography: Henri Lefebvre in recent Anglo-American scholarship. *Antipode, 33*(5), 809–825. https://doi.org/10.1111/1467-8330.00218

Elden, S. (2004). Between marx and heidegger: Politics, philosophy and Lefebvre's the production of space. *Antipode, 36*(1), 86–105. https://doi.org/10.1111/j.1467-8330.2004.00383.x

Elden, S. (2014). *Understanding Henri Lefebvre*. Igarss 2014. London and New York: Continuum. https://doi.org/10.1007/s13398-014-0173-7.2

Eliash, H., & Moreno, M. (1989). *Arquitectura y modernidad en Chile, 1925–1965: una realidad múltiple*. Ediciones Universidad Católica de Chile.

Engel, E. (2016). *Concesiones sesgadas, por Eduardo Engel: Voces de LA TERCERA*. Retrieved November 16, 2017, from www.latercera.com/voces/concesiones-sesgadas/

Espinoza, V. (1988). *Para una historia de los pobres en la ciudad*. Santiago: Ediciones SUR.

Fernandes, E. (2007). Constructing the "right to the city" in Brazil. *Social & Legal Studies, 16*(2), 201–219. https://doi.org/10.1177/0964663907076529

Fernandez, R. (2016). Financialization and housing: Between globalization and varieties of capitalism. In Albers, M. The Financialization of Housing (pp. 81–100). London-New York: Routledge.

Fezer, J. (2013). *Design in & against the neoliberal city*. London: Bedford Press.

Follain, J. R., & Giertz, S. H. (2013). Preventing House Price Bubbles: Lessons from the 2006–2012 Bust. New York: Lincoln Institute of Land Policy.

Foroughmand Araabi, H. (2017, November). Schools and skills of critical thinking for urban design. *Journal of Urban Design, 4809*, 1–17. https://doi.org/10.1080/13574809.2017.1369874

Fossa, L. (2011). *Contraloría sancionó a empleados municipales por irregularidades en permisos de edificación de Costanera Center*. Retrieved May 14, 2016, from http://ciperchile.cl/2011/04/14/contraloria-sanciono-a-empleados-municipales-por-irregularidades-en-permisos-de-edificacion-de-costanera-center/

Francese, J. (2009). *Perspectives on gramsci: Politics, culture and social theory*. London and New York: Routledge. Retrieved from http://books.google.com/boo ks?id=DnB4AgAAQBAJ&pg=PR5&dq=intitle:Perspectives+on+Gramsci+Poli tics+Culture+and+Social+Theory&hl=&cd=1&source=gbs_api%5Cnpapers3:// publication/uuid/6E3DF376-F662-4845-9889-E2DA75234B5E

Frey, R. G., & Wellman, C. H. (2003). *A companion to applied ethics*. Oxford, Malden, and Melbourne: Blackwell.

Friedman, M. (1987, Jun 24). Please reread your Adam Smith. *Wall Street Journal*, 26.

138   *Bibliography*

Friedman, M., & Pinochet, A. (1975). *Friedman-Pinochet Correspondence.pdf*. Retrieved from http://wwww.naomiklein.org/files/resources/pdfs/friedman-pinochet-letters.pdf

Friedmann, J. (2000). The good city: In defense of utopian thinking. *International Journal of Urban & Regional Research*, *24*(2), 460–472. https://doi.org/10.2166/wp.2008.061

Fubini, E. (2010). *La Estética Musical desde la Antigüedad hasta el siglo XX*. Madrid: Alianza Editorial.

Fuentes, C., & Valdebenito, R. (2015). *Chicago Boys*. Chile: La Ventana Cine.

Galetovic, A., & Poduje, I. (2006). *Santiago ¿Donde estamos y hacia donde vamos?* Santiago: Centro de Estudios Publicos CEP.

Gandy, M. (2011). *Urban constellations. Urban constellations*. Retrieved from http://simsrad.net.ocs.mq.edu.au/login?url=http://search.proquest.com/docview/1221844532?accountid=12219%5Cnhttp://multisearch.mq.edu.au/openurl/61MACQUARIE_INST/MQ_SERVICES_PAGE?url_ver=Z39.88-2004&rft_val_fmt=info:ofi/fmt:kev:mtx:book&genre=book&sid=Pro

Garnier, J. P. (1994). La vision urbaine de Henri Lefebvre. *Espaces et sociétés*, (2), 123–148.

Garreton, M. (2014). Derecho a la ciudad y participación frente al centralismo en Chile. *Revista*, *180*(34), 4–9.

Garreton, M. (2017). City profile: actually existing neoliberalism in Greater Santiago. *Cities*, 65, 32–50.

Gasic Klett, I. (2016). *Efecto Caval : una lupa sobre las sociedades y capitales que operan en el mercado del suelo*. Retrieved January 1, 2017, from http://ciperchile.cl/2016/12/22/efecto-caval-una-lupa-sobre-las-sociedades-y-capitales-que-operan-en-el-mercado-del-suelo/

Gatica, Y. C. (2011). La recuperación urbana y residencial del centro de Santiago: Nuevos habitantes, cambios socioespaciales significativos. *Eure*, *37*(112), 89–113. https://doi.org/10.4067/S0250-71612011000300005

Gay, C. (1854). *Atlas de la historia física y política de Chile*. Paris: Imprenta de E. Thunot.

Gehl, J. (2010). *Cities for people*. Washington, DC: Island Press.

George, R. V. (1997). A procedural explanation for contemporary urban design. *Journal of Urban Design*, *2*(2), 143–161. https://doi.org/10.1080/13574809708724401

Gilbert, A. (2001). La vivienda en América Latina. Washington: Banco Interamericano de Desarrollo.

Glaeser, E. L., & Meyer, J. R. (2002). *Chile: Political economy of urban development*. Cambridge, MA: Harvard University Press.

Gobierno de Chile. (2014). *Política Nacional de Desarrollo Urbano*. Santiago: Consejo Nacional de Desaarrollo Urbano.

Godoy Urzúa, H. (1981). *El Carácter chileno: estudio preliminar y selección de ensayos por Hernán Godoy Urzúa*. Santiago: Editorial Universitaria.

Golany, G. (1995). *Ethics and urban design: Culture, form, and environment*. New York, Chichester, Brisbane, Toronto, and Singapore: Wiley. Retrieved from http://media.wiley.com/spa_assets/R16B096RC3/site/shared/include/static/images/google_preview.gif

Gomez, M. (2017). *Santiago: South America's home for business*. Retrieved May 10, 2017, from www.businessdestinations.com/destinations/santiago-south-americas-home-for-business/

Gongora, M. (1960). *Origen de los inquilinos en el Chile central*. Santiago: Universidad de Chile.

Gonzalez Miranda, S. (2014). LAS INFLEXIONES DE INICIO Y TÉRMINO DEL CICLO DE EXPANSIÓN DEL SALITRE (1872–1919). UNA CRÍTICA AL NACIONALISMO METODOLÓGICO. *Dialogo Andino*, *1*(45), 39–50. Retrieved from www.scielo.cl/scielo.php?script=sci_arttext&pid=S0719-26812014000300005

Goonewardena, K. (2011). Henri Lefebvre y la revolución de la vida cotidiana, la ciudad y el Estado. *Urban*, (2), 1–15. Retrieved from file:///C:/Users/Marta/Downloads/Dialnet-HenriLefebvreYLaRevolucionDeLaVidaCotidianaLaCiuda-3762623.pdf

Goonewardena, K., Kipfer, S., Milgrom, R., & Schmid, C. (2008). *Space, difference, everyday life: Reading Henri Lefebvre* (Vol. 65). London and New York: Routledge. https://doi.org/10.1111/j.1745-7939.2009.01167_6.x

Gospodini, A. (2002). European cities in competition and the new'uses' of urban design. *Journal of Urban Design*, *7*(1), 59–73.

Grant, P., & Dube, R. (2015). A hard lesson on building in Chile: Ambitious plans for Santiago tower become quagmire for Chilean billionaire. *The Wall Street Journal*, p. n/a. Retrieved from https://search.proquest.com/docview/1701158716?accountid=14511

Grant, P., & Dube, R. (2015). A Hard Lesson on Building in Chile; Ambitious plans for Santiago tower become quagmire for Chilean billionaire. Retrieved from https://search.proquest.com/docview/1701158716?accountid=14511

Grazian, D. (2004). The right to the city: Social justice and the fight for public space. *Contemporary Sociology: A Journal of Reviews*, *33*(3), 361–362. https://doi.org/10.1177/009430610403300361

Greene, M., Rosas, J., & Valenzuela, L. (2011). *Santiago Proyecto Urbano*. Santiago: Ediciones ARQ.

Gross, P. (1990). Santiago de Chile: ideología y modelos urbanos. *EURE. Revista Latinoamericana de Estudios Urbano Regionales*, *16*(48), 67.

Gross, P. (1991). Santiago de Chile (1925–1990): planificación urbana y modelos políticos. *Revista EURE-Revista de Estudios Urbano Regionales*, *17*(52–53).

Guarda, G. (1978). *Historia Urbana del Reino de Chile*. Santiago: Editorial Andres Bello.

Gunder, M. (2011). Commentary: Is urban design still urban planning? An exploration and response. *Journal of Planning Education and Research*, *31*(2), 184–195. https://doi.org/10.1177/0739456X10393358

Gurovich, A. (2000). Conflictos y Negociaciones: La Planificación Urbana en el desarrollo del Gran Santiago, Chile. *Revista de Urbanismo*, *2*, 1–39. Retrieved from http://revistaurbanismo.uchile.cl/index.php/RU/article/viewArticle/12304

Harberger, A. (1979). Notas sobre los problemas de vivienda y planificación de la ciudad. *Revista AUCA*, *37*(1), 39–41.

Harries, K. (1997). *The Ethical Function of Architecture*. Cambridge: MIT Press.

Harvey, D. (2009). *Social Justice and the City*. Athens: University of Georgia Press.

Harvey, D. (1974). Class-monopoly rent, finance capital and the urban revolution. *Regional Studies*, *8*(3–4), 239–255. https://doi.org/10.1080/09595237400185251

Harvey, D. (1978). The urban process under capitalism: a framework for analysis. *International Journal of Urban and Regional Research, 2*(1–4), 101–131. https://doi.org/10.1111/j.1468-2427.1978.tb00738.x

Harvey, D. (1985). *The urbanization of capital.* Oxford: Blackwell.

Harvey, D. (2001). Globalization and the "Spatial Fix". *Geographische Revue: Marxism in Geography,* 23–30.

Harvey, D. (2005). *A brief history of neoliberalism.* New York: Oxford University Press.

Harvey, D. (2006). Neo-liberalism as creative destruction. *Geografiska Annaler, Series B: Human Geography, 88*(2), 145–158. https://doi.org/10.1111/j.0435-3684.2006.00211.x

Harvey, D. (2009). *Social justice and the city.* Athens, GA: University of Georgia Press.

Harvey, D. (2012). Ciudades rebeldes: Del derecho de la ciudad a la revolución urbana. Madrid: Ediciones Akal. https://doi.org/10.1007/s13398-014-0173-7.2

Harvey, D. (2014). *Seventeen contradictions and the end of capitalism.* London: Profile Books.

Herman, P. (2013). *El Minvu y la industria de la construcción.* Retrieved May 16, 2016, from www.defendamoslaciudad.cl/index.php/columnas/item/3770-el-minvu-y-la-industria-de-la-construcción

Herrera, F. (2015a). *Sobre la política de suelo e integración.* Santiago: CChC. Retrieved from www.cchc.cl/uploads/evento/archivos/02_FERNANDO_HERRERA_Propuesta paraunaPolitica.pdf

Herrera, F. (2015b). *Sobre la política de suelo e integración.* Santiago: CChC.

Hess, R., & Lefebvre, H. (1988). *Henri Lefebvre et l'aventure du siècle.* Editions Métailié.

Hidalgo-Dattwyler, R., Alvarado-Peterson, V., & Santana-Rivas, D. (2017). La espacialidad neoliberal de la producción de vivienda social en las áreas metropolitanas de Valparaíso y Santiago (1990–2014): ¿hacia la construcción idelógica de un rostro humano? *Cadernos Metrópole, 19*(39), 513–535. https://doi.org/10.1590/2236-9996.2017-3907

Hidalgo, R. (2011). NEGOCIOS INMOBILIARIOS Y LA TRANSFORMACIÓN\ rMETROPOLITANA DE SANTIAGO DE CHILE: DESDE LA RENOVACIÓN\ rDEL ESPACIO CENTRAL HASTA LA PERIFERIA EXPANDIDA, 1–16.

Holston, J. (2009). Insurgent citizenship in an era of global peripheries. *City and Society, 21*(2), 245–267. https://doi.org/10.1111/j.1548-744X.2009.01024.x.City

Holston, J., & Appadurai, A. (2008). Cities and Citizenship. In *State/Space* (pp. 296–308). Malden, MA, USA: Blackwell Publishing. https://doi.org/10.1002/9780470755686.ch17

Huchzermeyer, M. (2015, May). Reading Henri Lefebvre from the "global south": The legal dimension of his right to the city. *UHURU Seminar Series, Rhodes University,* 1–27.

Imilan, W., Olivera, P., & Beswick, J. (2016). Acceso a la vivienda en tiempos neoliberales: Un análisis comparativo de los efectos e impactos de la neoliberalización en las ciudades de Santiago, México y Londres. *Revista INVI. 31*(88). pp. 1–20. https://doi.org/10.4067/invi.v0i0.1093

Inam, A. (2002). Meaningful urban design: Teleological/catalytic/relevant. *Journal of Urban Design, 7*(1), 35–58. https://doi.org/10.1080/13574800220129222

Inostroza Córdova, L. I. (2015). Economía agroindustrial de Concepción y expansión triguera fronteriza: campesinos y mapuches en Biobío-Malleco, Chile, 1820–1850. *América Latina En La Historia Económica, 22*(1), 59–84.

Instituto Nacional de Estadísticas. (2017). Economía por Sectores Productivos. Retrieved January 1, 2018, from https://www.ine.cl/estadisticas/economicas

Inzulza-Contardo, J. (2011). Recuperando el derecho al espacio público desde la enseñanza de la arquitectura y el diseño urbano. *Revista De Arquitectura*, (24), 34–40. https://doi.org/10.5354/0716-8772.2011.26911

Inzulza-Contardo, J. (2016). Contemporary Latin American gentrification? Young urban professionals discovering historic neighbourhoods. *Urban Geography, 37*(8), 1195–1214. https://doi.org/10.1080/02723638.2016.1147754

Inzulza-Contardo, J., & Díaz Parra, I. (2016). Desastres naturales, destrucción creativa y gentrificación: estudio de casos comparados en Sevilla (España), Ciudad de México (México) y Talca (Chile). *Revista de Geografía Norte Grande, 128*(64), 109–128. https://doi.org/10.4067/S0718-34022016000200008

Jameson, F. (1985). Architecture and the critique of ideology. In J. Ockman (Ed.), *Architecture, criticism, ideology*. New Jersey: Princeton University Press.

Janoschka, M., & Hidalgo, R. (2014). *La Ciudad Neoliberal: estímulos de reflexión crítica*. Santiago: Editorial Universitaria.

Jenkins, P., Smith, H., & Wang, Y. P. (2006). Planning and housing in the rapidly urbanising world. *Planning and Housing in the Rapidly Urbanising World*, 1–368. https://doi.org/10.4324/9780203003992

Jiron, P., & Mansilla, P. (2014). Las consecuencias del urbanismo fragmentador en la vida cotidiana de habitantes de la ciudad de Santiago de Chile. *Eure, 40*(121), 79–97. https://doi.org/10.4067/S0250-71612014000300001

Johnson, C. G. (2011). The urban precariat, neoliberalization, and the soft power of humanitarian design. *Journal of Developing Societies, 27*(3–4), 445–475. https://doi.org/10.1177/0169796X1102700409

Jones, K. (2004). Book review of the urban revolution. *Journal of Regional Science, 44*(3), 589–638.

Kain, P. J. (1988). *Marx and ethics*. Oxford: Clarendon Press.

Kaysen, R. (2016). How much of my income should be budgeted for rent? *The New York Times*. Retrieved from https://nyti.ms/2ehXzWP REAL

Kaztman, R., & Retamoso, A. (2005). Segregación espacial, empleo y pobreza en Montevideo. *Revista de La CEPAL*, (85), 131–148.

Kipfer, S., & Goonewardena, K. (2013). Urban marxism and the post-colonial question: Henri Lefebvre and "colonisation"*. *Historical Materialism, 21*(2), 76–116. https://doi.org/10.1163/1569206X-12341297

Kipfer, S., Saberi, P., & Wieditz, T. (2012). Henri Lefebvre. *Handbuch Stadtsoziologie, 37*(1), 661–687. https://doi.org/10.1007/978-3-531-94112-7

Kipfer, S., Saberi, P., & Wieditz, T. (2013). Henri Lefebvre: Debates and controversies1. *Progress in Human Geography, 37*(1), 115–134. https://doi.org/10.1177/0309132512446718

Klein, N. (2010, March 3). Milton friedman did not save Chile. *The Guardian*. Retrieved from www.theguardian.com/commentisfree/cifamerica/2010/mar/03/chile-earthquake

Klein, N. (2011). The shock doctrine. *EuroEconomica*, 565. https://doi.org/10.1068/d2604ks

Klotz, M. (1993, May 8). "¿Donde se hace la arquitectura?" *Vivienda Y Decoracion*.

Kofman, E., & Lebas, E. (1996). *Henri Lefebvre: Writings on Cities*. Malden, MA: Blackwell. Malden, MA, USA: Blacwell.

Krätke, S. (2014). Cities in contemporary capitalism. *International Journal of Urban and Regional Research*, 38(5), 1660–1677.

Kuschnir, K. (2016). Ethnographic drawing: Eleven benefits of using a scketch-box for fieldwork. *Visual Ethnography*, 5(1), 103–134. https://doi.org/10.12835/VE2016.1-0060

Labra, M. E. (2002). La reinvención neoliberal de la inequidad en Chile: el caso de la salud. *Cadernos de Saúde Pública*, 18(4), 1041–1052. https://doi.org/10.1590/S0102-311X2002000400010

Lambiri, D., & Vargas, M. (2011). *Residential segregation and public housing policy, the case of Chile*, 1–47. Retrieved from http://udp.cl/descargas/facultades_carreras/economia/pdf/documentos_investigacion/wp29_Causas_Vargas.pdf

Lange, M., Mahoney, J., Hau, M., Lange, M., & Mahoney, J. (2006). Colonialism and development : Colonialism and development: A comparative analysis of Spanish and British colonies. *American Journal of Sociology*, 111(5), 1412–1462.

Langhorst, J. (2015). Rebranding the neoliberal city: Urban nature as spectacle, medium, and agency. *Architexture_Media Politics Society*, 6(4), 1–18. https://doi.org/10.14324/111.444.amps.2015v6i4.000

Latour, B. (2017). On Recalling Actors-Network Theory. *Philosophical Literary Journal Logos*, 27(1), 201–214. https://doi.org/10.22394/0869-5377-2017-1-201-214

Lawner, M. (1985). Un Terremoto indiscreto. *Cuadernos Del Instituto de Ciencias Alejandro Lipschutz*, 1, 12–18.

Lefebvre, H. (1958). Critique of everyday life volume 1. *Geografiska Annaler: Series B, Human Geography*, 77(1), 65. https://doi.org/10.2307/490375

Lefebvre, H. (1961). Utopie expérimentale: Pour un nouvel urbanisme. *Revue Française de Sociologie*, 2(3), 191–198. https://doi.org/10.2307/3319524

Lefebvre, H. (1970). *La révolution urbaine*. Paris: Gallimard.

Lefebvre, H. (1974). La producción del espacio. *Papers: Revista de Sociología*, 3, 219–229. https://doi.org/10.1017/CBO9781107415324.004

Lefebvre, H. (1975). What is the historical past? *New Left Review*, 27–34.

Lefebvre, H. (1980). Marxism exploded. *Review*, 4(1), 19–32. https://doi.org/10.2307/40240856

Lefebvre, H. (1991a). *Critique of everyday life* (Vol. 1). London and New York: Verso.

Lefebvre, H. (1991b). *The Production of Space*. Oxford and Cambridge, MA: Blackwell.

Lefebvre, H. (1996). *Writing on cities*. (E. Kofman & E. Lebas, Eds.). Oxford, MA: Blackwell.

Lefebvre, H. (2003). *The urban revolution*. Minneapolis: University of Minnesota Press.

Lefebvre, H. (2004). Elements of Rhythmanalysis: An introduction to the understanding rhythms. In *Rhythmanalysis: Space, time and everyday life*. London: Bloomsbury.

Lefebvre, H. (2009). *Dialectical materialism*. Minneapolis and London: University of Minnesota Press.

Lefebvre, H. (2014). *La producción del espacio*. Madrid: Captian Swing.

Lefebvre, H. (2016). *Marxist thought and the city* (1st ed.). Minneapolis: The University of Minnesota Press. Retrieved from www.upress.umn.edu/book-division/books/marxist-thought-and-the-city

Lefebvre, H., & Grindon, G. (2012). Revolutionary romanticism. *Art in Translation*, *4*(3), 287–300. https://doi.org/10.2752/175613112X13376070683270

Lopez, E., Jiron, P., Arriagada, C., & Eliash, H. (2014). *Chile Urbano hacia el Siglo XXI: Investigaciones y reflexiones de Política Urbana desde la Universidad de Chile*. (E. Lopez, P. Jiron, C. Arriagada, & H. Eliash, Eds.). Santiago: Editorial Universitaria.

López, E., & Meza, D. (2014). Neoliberalismo, regulación ad-hoc de suelo y gentrificación: el historial de la renovación urbana del sector Santa Isabel, Santiago. *Revista de Geografía Norte Grande*, *177*(58), 161–177. Retrieved from http://contested-cities.net/blog/neoliberalismo-regulacion-ad-hoc-de-suelo-y-gentrificacion-el-historial-de-la-renovacion-urbana-del-sector-santa-isabel-santiago/

López-Morales, E. J. (2009). Urban entrepreneurialism and creative destruction: A case-study of the urban renewal strategy in the peri-centre of Santiago de Chile, 1990–2005. *October*, 1990–2005. Retrieved from http://eprints.ucl.ac.uk/18707/

López-Morales, E. J. (2016). Gentrification in Santiago, Chile: A property-led process of dispossession and exclusion. *Urban Geography*, *37*(8), 1109–1131. https://doi.org/10.1080/02723638.2016.1149311

Low, S., & Smith, N. (Eds.). (2013). *The politics of public space*. London-New York: Routledge.

Lukács, G. (1980). *The ontology of social being: Labour*. London: Merlin.

Lukács, G. (2014). *Tactics and ethics 1919–1929*. London and New York: Verso.

Madanipour, A. (1997). Ambiguities of urban design. *The Town Planning Review*, *68*(3), 363–383. https://doi.org/10.2307/27798254

Madanipour, A. (2006). Roles and challenges of urban design. *Journal of Urban Design*, *11*(2), 173–193. https://doi.org/10.1080/13574800600644035

Maden, D., & Marcuse, P. (2008, December). In defense of housing. *Philosophical Papers*, *47*(2011), 85–96. https://doi.org/10.1300/J159v03n01

Maden, D., & Marcuse, P. (2016). *In defense of housing*. London and New York: Verso. https://doi.org/10.1300/J159v03n01

Marcuse, P. (2009). From critical urban theory to the right to the city. *City*, *13*(2–3), 185–197. https://doi.org/10.1080/13604810902982177

Marcuse, P. (2010). In defense of theory in practice. *City*, *14*(1), 4–12. https://doi.org/10.1080/13604810903529126

Marshall, M. N. (1996). Sampling for qualitative research. *Family Practice*, *13*(6), 522–525. https://doi.org/10.1093/fampra/13.6.522

Martin, J. Y. (2006). Une géographie critique de l'espace du quotidien. L'actualité mondialisée de la pensée spatiale d'Henri Lefebvre. *Journal of Urban Research*, (2).

Martin, P. Y., & Turner, B. A. (1986). Grounded theory and organizational research. *The Journal of Applied Behavioral Science*, *22*(2), 141–157. https://doi.org/10.1177/002188638602200207

Martínez, C. F., Hodgson, F., Mullen, C., & Timms, P. (2017, September). Creating inequality in accessibility: The relationships between public transport and social housing policy in deprived areas of Santiago de Chile. *Journal of Transport Geography*, 0–1. https://doi.org/10.1016/j.jtrangeo.2017.09.006

Martínez, R. (2003). The classical model of the Spanish-American colonial city. *The Journal of Architecture*, *8*(3), 355–368. https://doi.org/10.1080/13602360 32000134844

Martínez Lemoine, R. (2003). The classical model of the Spanish-American colonial city. *The Journal of Architecture*, *8*(3), 355–368. https://doi.org/10.1080/ 1360236032000134844

Martinez Lemoine, R. (2007). Santiago, los planos de transformación. 1984–1929. *Diseño Urbano Y Paisaje*, *11*(4), 14.

Marx, C., Engels, F., Lowy, M., & Kohan, N. (2013). El Manifiesto Comunista. *Editorial Laura/Lecturas Proletarias*, 1–48.

Marx, K. (1976). *Capital* (Vol. 1). London and New York: Penguin Books.

Marx, K. (1994). *Early writings*. Penguin Books. Retrieved from https://libcom.org/ files/Marx-Early-Writings.pdf

Mason, P. (2015). *Postcapitalism: A guid to our future*. London: Penguin Books.

May, T. (2001). *Social research: Issues, Methods and process: Social research.* Retrieved from www.amazon.co.uk/dp/0335206123

Mayol, A., Azocar, C., & Azocar, C. (2012). *El Chile Profundo*. Santiago: Editorial Universitaria.

Mayorga Henao, J. (2017). *Segregación residencial e inequidad en el acceso a servicios colectivos de educación, recreación y cultura en Bogotá-Colombia.* Retrieved from http://repositorio.uchile.cl/handle/2250/143772

McGuirk, J. (2014). *Radical cities*. New York and London: Verso.

Merrifield, A. (1993). Place and space: A Lefebvrian reconciliation. *Transactions of the Institute of British Geographers*, *18*(4), 516–531. https://doi.org/10. 2307/622564

Merrifield, A. (2002). *Metromarxism: A Marxist Tale of the city.* https://doi.org/ 10.1888/0333750888/6009

Merrifield, A. (2006a). *Henri Leferbvre: A critical introduction: Chemistry &. . . .* Retrieved from http://onlinelibrary.wiley.com/doi/10.1002/cbdv.200490137/ abstract

Merrifield, A. (2006b). *Henri Leferbvre: A critical introduction: Chemistry &. . . .* Retrieved from http://onlinelibrary.wiley.com/doi/10.1002/cbdv.200490137/abstract

Merrifield, A. (2011). The right to the city and beyond. *City*, *15*(3–4), 473–481. https://doi.org/10.1080/13604813.2011.595116

MIDSEO. (2013). *Metodología General del Sistema Nacional de Inversión*. Santiago: Sistema Nacional de Inversion Social - MIDSEO.

MINVU. (2014). *Vivienda Social En Copropiedad*. Santiago: Ministerio de Vivienda y Urbanismo.

MINVU. (2017). *Ley General de Urbanismo y Construcciones*. Santiago: Ministerio de Vivienda y Urbanismo.

Mitchell, D. (1995). The end of public space? People's park, definitions of the public, and democracy. *Annals of the Association of American Geographers*, *85*(1), 108–133. https://doi.org/10.1111/j.1467-8306.1995.tb01797.x

Monkkonen, P. (2012). La segregación residencial en el México urbano: Niveles y patrones. *Eure*, *38*(114), 125–146. https://doi.org/10.4067/S0250-716120120 00200005

Mont Pelerin Society, M. (2016). *Statement of aims*. Retrieved April 20, 2016, from www.montpelerin.org/

Moravanszky, Á., Schmid, C., & Stanek, L. (2014). *Urban revolution now: Henri Lefebvre in social research and architecture*. Farnham: Ashgate.

Mouton, J., & Marais, H. (2010). *Basic Concepts in the methodology of the social sciences*. https://doi.org/10.1002/9780470925409.ch1

Munizaga, G. (1980). Cronología sobre urbanismo y diseño urbano en Chile, 1870–1970. *EURE: Revista Latinoamericana de Estudios Urbanos Y Territoriales*, *6*(18), 69–90.

Munizaga, G. (2014). *Diseño urbano. Teoría y método*. Santiago: Ediciones UC.

Navarro, M. J. N. (2004). JEREMY BENTHAM Y EL LIBERALISMO EN CHILE DURANTE LA PRIMERA MITAD DEL SIGLO XIX1. *Boletín de La Academia Chilena de La Historia*, *70*(113), 285–313.

Oc, T. (2014). Reflections on urban design. *Journal of Urban Design*, *19*(1), 1. https://doi.org/10.1080/13574809.2013.863463

OECD. (2013). *OECD Urban Policy Reviews*, Chile 2013. Paris: OECD Publishing. https://doi.org/10.1787/9789264191808-en

Ostrom, E. (1996). Crossing the great divide: Coproduction, synergy, and development. *World Development*, *24*(6), 1073–1087.

Owen, R. (2011). A new view of society. In Capaldi, N., & Lloyd, G. *The Two Narratives of Political Economy*. Hoboken, NJ: John Wiley & Sons, pp. 229–240. https://doi.org/10.1002/9781118011690

Paget-Seekins, L. (2015). Bus rapid transit as a neoliberal contradiction. *Journal of Transport Geography*, *48*, 115–120. https://doi.org/10.1016/j.jtrangeo.2015.08.015

Peck, J. (2008). Remaking laissez-faire. *Progress in Human Geography*, *32*(1), 3–43. https://doi.org/10.1177/0309132507084816

Peck, J., Theodore, N., & Brenner, N. (2013). Neoliberal urbanism redux?. *International Journal of Urban and Regional Research*, *37*(3), 1091–1099.

Peck, J., Theodore, N., & Brenner, N. (2009). Neoliberal urbanism: Models, moments, mutations. *SAIS Review of International Affairs*, *29*(1), 49–66.

Perez, F. (2016). *Arquitectura en el Chile del siglo XX*. Santiago: Ediciones ARQ.

Pérez Oyarzun, F., Rosas, J., & Valenzuela, L. (2005). Las aguas del centenario. *ARQ (Santiago)*, (60), 72–74.

Peters, M. A. (2007). Foucault, biopolitics and the birth of neoliberalism. *Critical Studies in Education*, *48*(2), 165–178. https://doi.org/10.1080/17508480701494218

Piketty, T. (2014). *Capital in the XXI Century*. Boston, MA: Harvard University Press.

Poduje, I. (2012). Crecimiento urbano y vivienda social. Retrieved May 1, 2018, from http://95propuestas.cl/site/wp-content/uploads/2013/05/crecimiento-urbano-y-vivienda-social-ivan-poduje.pdf

Poduje, I., Jobet, N., & Martinez, J. (2015). *Infilling: cómo cambio Santiago y nuestra forma de vivir la ciu- dad*. Santiago: Atisba-Inmobiliaria Almagro.

Punter, J. (2010). Urban design and the English urban renaissance 1999–2009: A review and preliminary evaluation. *Journal of Urban Design*, *16*(1), 1–41. https://doi.org/10.1080/13574809.2011.521007

Purcell, M. (2013a). *The down-deep delight of democracy.* Oxford, MA and West Sussex: John Wiley & Sons.

Purcell, M. (2013b). To inhabit well: Counterhegemonic movements and the right to the city. *Urban Geography, 34*(4), 560–574. https://doi.org/10.1080/0272363 8.2013.790638

Racine, K. (2010). British cultural and intellectual influence in the Spanish American independence era. *Hispanic American Historical Review, 90*(3), 423–454.

Radio Cooperativa. (2012). Paulmann comparó Costanera Center con Torre Eiffel: Es bellísimo. Retrieved February 11, 2018, from https://www.cooperativa.cl/noticias/economia/empresas/grupos-economicos/paulmann-comparo-costanera-center-con-torre-eiffel-es-bellisimo/2012-04-24/173420.html

Rafferty, D. T. (2017). A critique of Jersey city, NJ's neoliberal, trickle-down economic ideology and an alternative development strategy. *Theory in Action, 10*(2), 1–21. https://doi.org/10.3798/tia.1937-0237.1708

Roberts, M. (2016). *The long depression.* Chicago: Haymarket Books.

Rodriguez, A., & Rodriguez, P. (2009). *Santiago, una ciudad neoliberal.* Quito: OLACCHI.

Rodríguez, A., & Sugranyes, A. (2005). *Los con techo: un desafío para la política de vivienda social.* Santiago: Ediciones SUR.

Rodríguez Weber, J. (2009). *Los tiempos de la desigualdad. La distribución del ingreso en Chile, entre la larga duración, la globalización y la expansión de la frontera, 1860–1930.* Universidad de la Republica. Retrieved from www.academia.edu/419660/Los_tiempos_de_la_desigualdad._La_distribucion_del_ingreso_en_Chile_entre_la_larga_duracion_la_globalizacion_y_la_expansion_de_la_frontera_1860-1930

Rolnik, R. (2013). Late neoliberalism: The financialization of homeownership and housing rights. *International Journal of Urban and Regional Research, 37*(3), 1058–1066. https://doi.org/10.1111/1468-2427.12062

Romero, H., Irarrázaval, F., Opazo, D., Salgado, M., & Smith, P. (2010). Climas urbanos y contaminación atmosférica en santiago de chile hugo romero. *Eure, 36*(109), 35–62. https://doi.org/10.4067/S0250-71612010000300002

Rosas, J., Hidalgo, G., Strabucchi, W., & Bannen, P. (2015). El Plano Oficial de Urbanización de la Comuna de Santiago de 1939: Trazas comunes entre la ciudad moderna y la ciudad preexistente. *ARQ (Santiago),* (91), 82–93. https://doi.org/10.4067/S0717-69962015000300013

Rossi, U., & Vanolo, A. (2015). Urban Neoliberalism. In J. Wright (Ed.), *International Encyclopedia of the Social & Behavioral Sciences* (pp. 846–853). London: Elsevier. https://doi.org/10.1016/B978-0-08-097086-8.74020-7

Ruiz, C., & Boccardo, G. (2014). *Los chilenos bajo el neoliberalismo.* Santiago: El Desconcierto.

Saad-Filho, A., & Fine, B. (2004). Marx's capital. London: Pluto Press.挴

Sabatini, F. (2000). Reforma de los mercados de suelo en Santiago, Chile: efectos sobre los precios de la tierra y la segregación residencial. *Revista Latinoamericana de Estudios Urbano Regionales, 23*(77), 49–80. https://doi.org/10.4067/S0250-71612000007700003

Sabatini, F., & Brain, I. (2008). La segregacion, los guetos y la integracion social urbana: Mitos y claves. *Eure, 34*(103), 5–26. https://doi.org/10.4067/S0250-71612008000300001

Sabatini, F., Wormald, G., & Rasse, A. (2013). *Segregación de la vivienda social: ocho conjuntos en Santiago, Concepción y Talca.* Santiago: Instituto de Estudios Urbanos y Territoriales, Facultad de Arquitectura, Diseño y Estudios Urbanos, Pontificia Universidad Católica de Chile.

Salas, V. (2000). *Rasgos históricos del movimiento de pobladores en los últimos 30 años* (Vol. 32). Retrieved from recuperado en www. accioncultural.cl/html/pdf/ movimiento_pobladores. pdf

Salazar, G. (2003a). *Historia de la acumulación capitalista en Chile (apuntes de clases).* Santiago: LOM Ediciones.

Salazar, G. (2003b). *La Historia Desde Abajo Y Desde Dentro,* 476. Retrieved from www.bncatalogo.cl/F/CYPDYBUFDH624UCCDUEMDETLP4J98K5PNR 565FVT523EKLNCBE-46736?func=full-set-set&set_number=782477&set_ entry=000001&format=040

Salazar, G. (2009). Mercaderes, empresarios y capitalistas (Chile, siglo XIX). Santiago: Sudamericana.

Salazar, G. (2012a). *Clase Magistral – Gabriel Salazar: Padres y Madres de la Patria – YouTube.* Retrieved November 16, 2017, from www.youtube.com/ watch?v=8J2IHjm8muM

Salazar, G. (2012b). *Movimientos sociales en Chile: trayectoria histórica y proyección política* (Vol. 3). Santiago: Uqbar.

Salazar, G. (2015). *La enervante levedad historica de la clase politica civil (Chile 1900–1973).* Santiago: Debate.

Salerno, B. (2014). Neoliberalismo En Buenos Aires. *La Ciudad Neoliberal,* 129–149.

Sassen, S. (1991). *The global city.* New York. New Jersey: Princeton University Press.

Scharpf, F. W. M. (1997). *Games real actors play: Actor-centered Institutionalism in policy research.* London: Westview.

Schipper, S. (2014). The financial crisis and the hegemony of urban neoliberalism: Lessons from Frankfurt am main. *International Journal of Urban and Regional Research, 38*(1). https://doi.org/10.1111/1468-2427.12099

Shaw, K. (2015). Planetary urbanisation: what does it matter for politics or practice?. *Planning Theory & Practice, 16*(4), 588–593.

Shields, R. (1989). Social spatialization and the built environment: The West Edmonton Mall. *Environment & Planning D: Society & Space, 7*(2), 147–164. https://doi.org/10.1068/d070147

Shields, R. (2005). *Lefebvre: Love and struggle: Spatial dialectics* (Vol. 53). London and New York: Routledge. https://doi.org/10.1017/CBO9781107415324.004

Short, J. R. (2006). *Urban theory: A critical asssessment.* London: Palgrave Macmillan.

Silbiger, S. (2009). *The ten-day MBA.* New York and London: HarperCollins Publishers.

Silva Vargas, F. (1963). Tierras y pueblos de indios en el Reino de Chile. Esquema historico-jurídico. *Director,* 253.

Simian, J. M. (2010). Logros e Desafíos de La Política Habitacional en Chile. *Estudios Públicos,* (117), 270–322.

Singleton, T. A. (2001). Slavery and spatial dialectics on Cuban coffee plantations. *World Archaeology, 33*(1), 98–114.

Sistema Nacional de Inversiones. (2017). *Precios Sociales 2017.* Santiago: Ministerio de Desarrollo Social.

Sklair, L. (2005). The transnational capitalist class and contemporary architecture in globalizing cities. *International Journal of Urban and Regional Research, 29*(3), 485–500.

Sklair, L. (2006). Iconic architecture and capitalist globalization. *City, 10*(1), 21–47.

Slater, T. (2017). Planetary rent gaps. *Antipode, 49*, 114–137. https://doi.org/10.1111/anti.12185

Smith, A. (1776). An inquiry into the wealth of nations. *Strahan and Cadell, London*, 1–11. https://doi.org/10.7208/chicago/9780226763750.001.0001

Smith, N. (2009). Toxic capitalism. *New Political Economy, 14*(3), 407–412.

Smith, N. (2011). Uneven development redux. *New Political Economy, 16*(2), 261–265. https://doi.org/10.1080/13563467.2011.542804

Smith, P., & Max-Neef, M. (2011). *Economics unmasked: From power and greed to compassion and the common good.* Cambridge: UIT Cambridge Limited.

Soja, E. W. (1980). The socio-spatial dialectics. *Annals of the Association of American Geographers, 70*(2), 207–225. https://doi.org/doi:10.1111/j.1467-8306.1980.tb01308.x

Solimano, A. (2012). *Chile and the neoliberal trap: The post-Pinochet era.* Cambridge: Cambridge University Press.

Solimano, A. (2014). Neoliberalismo y desarrollo desigual: La experiencia chilena. *Revista Nueva Economía Sustentable, 1*(1), 1–19

Sorkin, M. (2013). *All over the map: Writing on buildings and cities.* London and New York: Verso.

Soublette, G. (2015). *Entrevista a Gastón Soublette – Parte III: La Cultura Mapuche.* Retrieved February 12, 2016, from www.youtube.com/watch?v=N27LAd906yM

Spencer, D. (2016). *The architecture of neoliberalism.* London: Bloomsbury.

Springer, S. (2016). *The handbook of neoliberalism.* London: Routledge.

Stanek, Ł. (2008a). Methodologies and situations of urban research: Re-reading Henri Lefebvre's. *The Production of Space, 4*, 461–465.

Stanek, L. (2008b). Space as concrete abstraction. *Space, Difference, Everyday Life: Reading Lefebvre*, 62–79. Retrieved from www.henrilefebvre.org/text/Routledge_STANEK.pdf

Stanek, Ł. (2011). *Henri Lefebvre on space* (Vol. 1). Minneapolis and London: University of Minnesota Press. https://doi.org/10.1017/CBO9781107415324.004

Stanek, Ł. (2017, June). Collective luxury. *Journal of Architecture, 2365.* https://doi.org/10.1080/13602365.2017.1307871

Stanek, T., & Schmid, C. (2011). Teoría, no método: Henri Lefebvre, investigación y diseño urbanos en la actualidad. *Urban*, 02, 59–66.

Stiglitz, J. E. (2010). Freefall: America, free markets, and the sinking of the world economy. *WW Norton & Company*, 258. https://doi.org/10.1111/j.1468-2346.2008.00711.x

Stiglitz, J. E. (2017, June 7). Austerity has strangled the British economy: Only Labour gets this | Joseph Stiglitz | Opinion | The Guardian. *The Guardian.* Retrieved from www.theguardian.com/commentisfree/2017/jun/07/austerity-britain-labour-neoliberalism-reagan-thatcher

Stillerman, J. (2016). Educar a niñas y niños de clase media en Santiago: Capital cultural y segregación socioterritorial en la formación de mercados locales de educación. *Eure, 42*(126), 169–186. https://doi.org/10.4067/S0250-71612016000200008

Storper, M. (2015, August). The neoliberal city as idea and reality. *Territory, Politics, Governance, 2671*, 1–37. https://doi.org/10.1080/21622671.2016.1158662

Subsecretaría de Redes Asistenciales, & Ministerio de Salud. (2016). Informe sobre Brechas de Personal de Salud por Servicio de Salud. *Glosa 01 Letra I. Ley de Presupuesto No 20.882 Y 20.890 Año 2016*, (19), 71. Retrieved from http://web.minsal.cl/wp-content/uploads/2015/08/Informe-Brechas-RHS-en-Sector-Público_Abril2017.pdf%0Ahttp://web.minsal.cl/wp-content/uploads/2015/08/Informe-Brechas-RHS-en-Sector-Público_Marzo2016.pdf

Swyngedouw, E., & Heynen, N. C. (2003). Urban political ecology, justice and the politics of scale. *Antipode, 35*(5), 898–918. https://doi.org/10.1111/j.1467-8330.2003.00364.x

Taşan-Kok, T., & Baeten, G. (Eds.). (2012). *Contradictions of neoliberal urban planning: Contradictions of neoliberal urban planning.* London and New York: Springer. https://doi.org/10.1007/978-90-481-8924-3_1

Thayer Ojeda, T. (1905). *Santiago durante el siglo XVI: constitución de la propiedad urbana i noticias biográficas de sus primeros pobladores.* Santiago: Imprenta Cervantes.

Theodore, N., Peck, J., & Brenner, N. (2009). Urbanismo neoliberal: la ciudad y el imperio de los mercados. *Temas Sociales SUR,* 12. Retrieved from www.sitiosur.cl

Theodore, N., Peck, J., & Brenner, N. (2012). Neoliberal urbanism: Cities and the rule of markets. *The New Blackwell Companion to the City,* 15–25. https://doi.org/10.1002/9781444395105.ch2

Thomas, G. (2007). *Education and theory: Strangers in paradigms.* Berkshire: McGraw Hill.

Tickell, A., & Peck, J. (2002). Neoliberalizing space. *Antipode, 34*(3), 380–404. https://doi.org/10.1111/1467-8330.00247

Torrejón, F. (2001). Variables geohistóricas en la evolución del sistema económico pehuenche durante el periodo colonial. *Revista Universum, 16,* 219–236.

Tozzi, L. (2016). Speciale Biennale Architettura 2016. AlfaBeta2, 1–10. Retrieved from https://www.alfabeta2.it/2016/06/11/larchitettura-la-differenza/

Trivelli, P. (1981). Reflexiones en torno a la política Nacional de Desarrollo Urbano. *EURE: Revista Latinoamericana de Estudios Urbanos Y Territoriales, 8*(22), 43–64.

Ureta, S. (2014). Normalizing transantiago: On the challenges (and limits) of repairing infrastructures. *Social Studies of Science, 44*(3), 368–392. https://doi.org/10.1177/0306312714523855

Valencia, M. (2007). Revolución neoliberal y crisis del Estado Planificador. El desmontaje de la planeación urbana en Chile. 1975–1985. *Diseño Urbano Y Paisaje, 4*(12), 1–25.

Valenzuela, E., Penaglia, F., & Basaure, L. (2016). Acciones colectivas territoriales en Chile, 2011–2013: De lo ambiental-reivindicativo al autonomismo regionalista. *Eure, 42*(125), 225–250. https://doi.org/10.4067/S0250-71612016000100010

van Ham, M., Tammaru, T., de Vuijst, E., & Zwiers, M. (2016). Spatial segregation and socio-economic mobility in European cities. *IZA Discussion Paper Series,* (IZA DP No. 10277). https://doi.org/10.13140/RG.2.2.19934.43841

Vargas, M. (2006). *Causes of Residential Segregation: The Case of Santiago, Chile.* Working Paper Series of Centre for Spatial and Real Estate Economics, Department of Economics, University of Reading.

Vargas, M. (2016). Tacit collusion in housing markets: The case of Santiago, Chile. *Applied Economics, 48*(54), 5257–5275. https://doi.org/10.1080/00036846.2016.1176111

Vergara, J. E. (2017). VERTICALIZACIÓN. LA EDIFICACIÓN EN ALTURA EN LA REGIÓN METROPOLITANA DE SANTIAGO (1990–2014)*. *Revista INVI*, *32*(90), 9–49.

Vergara-Perucich, F. (2011). *El Espacio Cívico. El Paseo Bulnes como caso emblemático*. Pontificia Universidad Catolica de Chile.

Vicuña, M. (2013, August). El marco regulatorio en el contexto de la gestión empresarialista y la mercantilización del desarrollo urbano the regulatory framework within the context of the business administration and the commodification of urban development in. *2Revista INVI*, *28*, 181–219.

Vicuña Mackenna, B. (1869). *Historia critica y social de la ciudad de Santiago desde su fundacion hasta nuestros dias (1541–1868)*. Valparaiso: Imprenta del Mercurio.

Vicuña Mackenna, B. (1872). La transformación de Santiago. Notas e indicaciones respetuosamente sometidas a la Ilustre Municipalidad, al Supremo Gobierno y al Congreso Nacional por el Intendente de Santiago. Santiago, Chile: Imprenta de la Librería del Mercurio.

Villaça, F. (2011). São Paulo: segregação urbana e desigualdade. *Estudos Avançados*, *25*(71), 37–58. https://doi.org/10.1590/S0103-40142011000100004

Vitale, L. (2011). *Interpretacion Marxista de la Historia de Chile. Interpretacion Marxista*. Santiago: LOM Ediciones. https://doi.org/10.1017/CBO9781107415324.004

Weber, R. (2002). Extracting value form the city: Neoliberalism and urban redevelopment: ZBZ Libraries and Consortia ZAD50. *Antipode*, *34*(3), 519–540.

Whitehead, M. (2003). Love thy neighbourhood: Rethinking the politics of scale and Walsall's struggle for neighbourhood democracy. *Environment and Planning A*, *35*(2), 277–300.

Wisdom, J., & Creswell, J. W. (2013). Integrating quantitative and qualitative data collection and analysis while studying patient-centered medical home models. *Agency for Healthcare Reseach and Quality*, (13-0028-EF), 1–5. https://doi.org/No.13-0028-EF

World Bank (2016). *World Bank open data*. Retrieved from https://data.worldbank.org

Yañez, M. (2014). Evolución en el nivel de satisfacción de las personas vulnerables de Santiago de Chile con el Transantiago, años 2007 a 2014. *OIKOS*, *18*(38), 127–163.

Yin, R. K. (2009). *Case study research and applications: Design and methods*. London: Sage publications.

Zapater, H. (1997). Huincas y mapuches. *Historia1*, *1*(30), 441–504.

Zimbalist, A., & Sherman, H. J. (1984). *Comparing economic systems: Comparing economic systems*. Orlando and London: Academic Press, Inc. https://doi.org/10.1016/B978-0-12-781050-8.50023-3

Žižek, S. (2008). The secret clauses of the liberal Utopia. *Law and Critique*, *19*(1), 1–18. https://doi.org/10.1007/s10978-007-9023-5

Zizek, S. (2012). Introduction: The spectre of ideology. *Mapping Ideology*, 1–33. https://doi.org/10.1016/S0074-7696(08)61919-1

Zizek, S. (2015). *Slavoj Žižek: Democracy and capitalism are destined to split up: Video | Big Think*. Retrieved April 7, 2017, from http://bigthink.com/videos/slavoj-zizek-on-capitalism-and-the-commons

Zunino, H. M. (2006). Power relations in urban decision-making: Neo-liberalism, "techno-politicians" and authoritarian redevelopment in Santiago, Chile. *Urban Studies*, *43*(10), 1825–1846. https://doi.org/10.1080/00420980600838184

Index

Note: Page numbers in *italics* indicate figures and page numbers in **bold** indicate tables.